W9-AQM-007

**NORTH
STAR
WAY**

Solve
for
Happy

ENGINEERING YOUR
PATH TO JOY

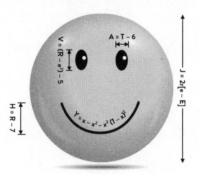

Mo Gawdat

NORTH
STAR
WAY

New York London Toronto Sydney New Delhi

NORTH
STAR
WAY

An Imprint of Simon & Schuster, Inc.
1230 Avenue of the Americas
New York, NY 10020

Copyright © 2017 by Mo Gawdat

All rights reserved, including the right to reproduce this book or portions thereof
in any form whatsoever. For information address North Star Way Subsidiary
Rights Department, 1230 Avenue of the Americas, New York, NY 10020.

First North Star Way hardcover edition March 2017

NORTH STAR WAY and colophon are
trademarks of Simon & Schuster, Inc.

For information about special discounts for bulk purchases,
please contact Simon & Schuster Special Sales at
1-866-506-1949 or business@simonandschuster.com.

The North Star Way Speakers Bureau can bring authors to your live event.
For more information or to book an event contact the North Star Way Speakers
Bureau at 1-212-698-8888 or visit our website at www.thenorthstarway.com.

Interior design by Renato Stanisic

Manufactured in the United States of America

10 9 8 7 6 5 4 3 2

Library of Congress Cataloging-in-Publication Data is available.

ISBN 978-1-5011-5755-4
ISBN 978-1-5011-5759-2 (ebook)

The gravity of the battle means nothing to those at peace.

For Ali

I am sure you're happy wherever you are now

Contents

Introduction

Seventeen days after the death of my wonderful son, Ali, I began to write and couldn't stop. My topic was happiness—an unlikely subject given the circumstances.

Ali truly was an angel. He made everything he touched better and everyone he met happier. He was always peaceful, always happy. You couldn't miss his energy or how he affectionately cared for every being that ever crossed his path. When he left us, there was every reason to be unhappy—even miserable. So how did his departure lead me to write what you're about to read? Well, that's a story that started around the date of his birth—perhaps even earlier.

Since the day I started working, I have enjoyed a great deal of success, wealth, and recognition. Yet through it all, I was constantly unhappy. Early in my career with tech giants like IBM and Microsoft I gained an abundance of intellectual satisfaction, plenty of ego gratification, and, yes, I made a bit of money. But I found that the more fortune blessed me, the less happy I became.

This wasn't just because life had become complicated—you know, like that rap song from the 90s, "Mo Money mo Problems." The issue was that, despite the rewards both financial and intellectual, I was not able to find any joy in my life. Even my greatest blessing, my family, didn't give me the joy they might have because I didn't know how to receive it.

The irony was that as a younger man, despite the struggle to find my path in life and often just trying to make ends meet, I'd always been very happy. But by 1995, when my wife and I and our two children packed up and moved to Dubai, things had changed. Nothing against Dubai, mind you. It's a remarkable city whose generous citizens, the Emiratis, truly made us feel at home. Our arrival coincided with the breakout point of Dubai's explosive growth, which offered astounding career opportunities and millions of ways to make yourself happy, or at least try.

But Dubai can also feel surreal. Against a gleaming landscape of hot sand and turquoise water, the skyline is crowded with futuristic office buildings and residential towers where multimillion-dollar apartments are snapped up by a steady stream of global buyers. In the streets, Porsches and Ferraris jockey for parking spaces with Lamborghinis and Bentleys. The extravagance of the concentrated wealth dazzles you— but at the same time it tempts you to question whether, compared to all this, you've actually achieved much of anything.

By the time we arrived in the Emirates, I'd already fallen into the habit of comparing myself to my superrich friends and always coming up short. But those feelings of one-downs-man-ship didn't send me to the shrink or to the ashram. Instead it made me strive harder. I simply did what I'd always done as a geek who'd read obsessively since childhood: I bought a pile of books. I studied technical analyses of stock trends down to the basic equations that plotted every chart. And by learning them I could predict short-term fluctuations in the market like a pro. I would come home after finishing my day job at just about the time the NASDAQ opened in the United States and apply my math skills to making serious money as a day trader (or more accurately in my case, a night trader).

And yet—and I expect I'm not the first person you've heard tell this tale—the "mo' money" I made, the more miserable I became. Which led me to simply work harder and buy more toys on the misguided assumption that, sooner or later, all this effort was going to pay off and I'd find the pot of gold—happiness—thought to lie at the end of the high-achievement rainbow. I'd become a hamster on what psychologists call the "hedonic treadmill." The more you get, the more you want. The more you strive, the more reasons you discover for striving.

One evening I went online and with two clicks bought two vintage Rolls-Royces. Why? Because I could. And because I was desperately trying to fill the hole in my soul. You won't be surprised to hear that when those beautiful classics of English automotive styling arrived at the curb, they didn't lift my mood one bit.

Looking back at this phase in my life, I wasn't much fun to be around. My work was focused on expanding the business of Microsoft throughout Africa and the Middle East, which, as you might imagine, had me spending more time in airplanes than not. In my constant quest for *more* I'd become pushy and unpleasant even at home, and I knew it. I spent too little time appreciating the remarkable woman I'd married, too little time with my wonderful son and daughter, and never paused to enjoy each day as it unfolded.

Instead I spent most of my waking hours being driven, nervous, and critical, demanding achievement and performance even from my kids. I was manically trying to make the world conform to the way I thought it ought to be. By 2001 the relentless pace and the emptiness had led me into a very dark place.

At that point I knew I couldn't go on ignoring the problem. This pushy, unhappy person staring back at me in the mirror wasn't really me. I missed the happy, optimistic young man I'd always been, and I was tired of trudging along in this tired, miserable, aggressive-looking guy's shoes. I decided to take on my unhappiness as a challenge: I would apply my geek's approach to self-study, along with my engineer's analytical mind, to digging my way out.

Growing up in Cairo, Egypt, where my mother was a professor of English literature, I'd started devouring books long before my first day of school. Beginning at the age of eight, I chose a topic of focus each year and bought as many books as my budget could afford. I would spend the rest of the year learning every word in every book. This obsessiveness made me a joke to my friends, but the habit stuck with me as my approach to all challenges and ambitions. Whenever life turned tough, I read.

I went on to teach myself carpentry, mosaics, guitar, and German. I read up on special relativity, studied game theory and mathematics, and I learned to develop highly sophisticated computer programming. As a kid in grade school, and then as a teenager, I approached my piles of books with single-minded dedication. As I grew older, I applied that same passion for learning to classic car restoration, cooking, and hyperrealistic charcoal portraits. I achieved a reasonable level of proficiency in business, management, finance, economics, and investment mainly just from books.

When things get tough we tend to do more of what we know how to do best. So, in my thirties and miserable, I submerged myself in reading about my predicament. I bought every title I could find on the topic of happiness. I attended every lecture, watched every documentary, and then diligently analyzed everything I'd learned. But I didn't approach the subject from the same perspective as the psychologists who'd written the books and conducted the experiments that had made "happiness research" such a hot academic discipline. Certainly I didn't follow in the slipstream of all the philosophers and theologians who'd struggled with the problem of human happiness since civilization began.

In keeping with my training, I broke the problem of happiness down into its smallest components and applied an engineering analysis. I adopted a facts-driven approach that would be scalable and replicable. Along the way, I challenged every process I'd been told to blindly implement, tested the fit of every moving part, and looked deeply into the validity of every input as I worked to create an algorithm that would produce the desired result. As a software developer, I set a target to find the code that could be applied to my life again and again to predictably deliver happiness every time.

Oddly enough, after all this hyperrational effort worthy of Mr. Spock, I found my first real breakthrough during a casual conversation with my mother. She'd always told me to work hard and to prioritize my financial success above all. She frequently invoked an Arabic proverb that, loosely translated, meant "Eat frugally for a year and dress frugally for another, and you'll find happiness forever." As a young man I'd followed that advice religiously. I'd worked hard and saved and I'd become successful. I'd fulfilled my side of the bargain. So one day I went to ask my mom: Where was all that happiness I now had a right to expect?

During that conversation, it suddenly hit me that happiness shouldn't be something you wait for and work for as if it needs to be *earned*. Furthermore, it shouldn't depend on external conditions, much less circumstances as fickle and potentially fleeting as career success and rising net worth. My path till then had been full of progress and success, but every time I'd gained yardage on that field, it was as if they moved the goal posts back a little farther.

What I realized was that I would never get to happiness as long as I held on to the idea that as soon as I do this or get that or reach this benchmark I'll *become* happy.

In algebra, equations can be solved in many ways. If A=B+C, for example, then B=A−C. If you try to solve for A, you would look for the values of the other two parameters—B and C—and if you tried to solve for B, you would be taking different steps. The parameter you choose to solve for drastically changes your approach to the solution. The same is true when you decide to **solve for happy**.

I began to see that throughout all my striving I'd been trying to solve the wrong problem. I'd set myself the challenge of multiplying material wealth, fun, and status so that, *eventually*, the product of all that effort would be . . . happiness. What I really needed to do instead was to skip the intermediate steps and simply solve for happiness itself.

My journey took almost a decade, but by 2010 I'd developed an equation and a well-engineered, simple, and replicable model of happiness and how to sustain it that fit together perfectly.

I put the system to the test and it worked. Stress from losing a business deal, long security lines at the airport, bad customer service—none of it could dim my happiness. Daily life as a husband, parent, son, friend, and employee had its inevitable ups and downs, but no matter how any particular day went, good or bad—or a little of each—I found that I was able to enjoy the ride of the roller coaster itself.

I'd finally returned to being the happy person I recognized as the "me" when I first started out, and there I remained for quite a while. I shared my rigorous process with hundreds of friends, and my Happiness Equation worked for them as well. Their feedback helped me refine the model even further. Which, as it turned out, was a good thing, because I had no idea just how much I was going to need it.

My father was a distinguished civil engineer and an exceptionally kind man. Though my passion had always been computer science, I studied civil engineering just to please him. My field of study was not the biggest contribution to my education anyway because, as my father believed, learning takes place in the real world. Ever since I was in secondary school my father had encouraged me to spend each vacation in a different country. At first he squeezed every cent to make these experiences happen for me, and he made arrangements for me to visit with family or friends as I traveled. Later I worked to support the cost of my trips on my own. Those real-world experiences were so valuable that I vowed to offer a similar opportunity to my kids.

As luck would have it, my choice of university offered me the greatest benefit and blessing of those student days. I came to know a charming, intelligent woman named Nibal. A month after her graduation we married, and one year later she became Umm Ali, mother of Ali ☺, as women are called in the Middle East when their first child is born. Eighteen months after that, our daughter, Aya, came along to become the sunshine and the irrepressible, energizing force within our family. With Nibal, Ali, and Aya in my life, my good fortune knew no bounds. My love for my family drove me to work hard to provide the best life I could for them. I took on life's challenges like a charging rhino.

In 2007 I joined Google. Despite the company's success, its global reach was limited at that point, so my role was to expand our operations into Eastern Europe, the Middle East, and Africa. Six years later I moved over to Google X, now a separate entity known as X, where I eventually became the chief business officer. At X, we don't try to achieve incremental improvements in the way the world works; instead, we try to develop new technologies that will reinvent the way things are. Our goal is to deliver a radical, tenfold—10X—improvement. This leads us to work on seemingly sci-fi ideas such as autonomous carbon fiber kites to serve as airborne wind turbines, miniature computers built into contact lenses that capture physiological data and communicate wirelessly with other computers, and balloons to carry telecom technology into the stratosphere to provide Internet service to every human anywhere in the world. At X, we call these "moonshots."

When you're seeking modest improvement in what exists, you start working with the same tools and assumptions, the same mental framework on which the old technology is based. But when the challenge is to move ahead by a factor of ten, you start with a blank slate. When you commit to a moonshot, you fall in love with the problem, not the product. You commit to the mission before you even know that you have the ability to reach it. And you set audacious goals. The auto industry, for example, has been focused on safety for decades. They made consistent incremental progress by adding improvements to the traditional design of a car—the design we've all gotten used to since the early 1900s. Our approach at X is to begin by asking, "Why let an accident happen in the first place?" That's when we commit to the moonshot: a self-driving car.

Meanwhile, with my happiness model working well, and as I was deriving great pleasure from my career, doing my part to help invent the future, my son and daughter were learning and growing and, in keeping with my father's tradition, traveling to new places every summer. They had plenty of friends to visit across the globe, and they were always out exploring.

In 2014 Ali was a college student in Boston, and that year he had a long trip planned across North America, so we weren't expecting him to make it home to Dubai for his usual visit. I was pleasantly surprised, then, in May, when he called to say he felt an overwhelming desire to come and spend a few days with us. For some reason he felt a sense of urgency, and he asked if we could book him a flight home as soon as school was over. Aya was planning to visit too, so Nibal and I were happy beyond belief. We made the arrangements and eagerly looked forward to the joy of having the whole family together in July.

Four days after he arrived, Ali suffered an acute belly pain and was admitted to a local hospital, where the doctors prescribed a routine appendectomy. I wasn't concerned. In fact, I was relieved that this was happening while he was home so we could take care of him. The vacation might not have been going as I'd imagined, but the change in plans was easy enough to accommodate.

When Ali was on the operating table, a syringe was inserted to blow in carbon dioxide to expand his abdominal cavity and clear space for the rest of the procedure. But the needle was pushed just a few millimeters too far, puncturing Ali's femoral artery—one of the major vessels carrying blood from the heart. Then things went from bad to worse. Precious moments slipped by before anyone even realized the blunder, and then a series of additional mistakes were made with fatal consequence. Within a few hours, my beloved son was gone.

Before we could even begin to absorb the enormity of what had happened, Nibal, Aya, and I were surrounded by friends who helped us handle the practicalities and supported us while we struggled to comprehend the sharp turn our lives had just taken.

Losing a child, they say, is the hardest experience anyone can endure. It certainly shakes every parent to the core. Losing Ali at his prime was harder still, and losing him unexpectedly to preventable human error may have been the very hardest thing of all.

But for me, the loss was even worse because Ali was not just my son but also my best friend. He had been born when I was quite young, and I felt as if we'd grown up together. We played video games together, listened to music together, read books together, and laughed a lot together. At eighteen Ali was noticeably wiser than many men I knew. He was a support and a confidant. At times I even found myself thinking, **"When I grow up, I want to be just like Ali."**

Although all parents see their children as exceptional, I honestly believe that Ali truly was. When he left us, we received messages from all over the world, from hundreds of people who described how this twenty-one-year-old had changed their lives. Some of the people who wrote were in their teens, and some were well into their seventies. How Ali had found the time and wisdom to touch the lives of so many people, I'll never know. He was a role model for peacefulness, happiness, and kindness. And he had a sense of presence that spread those characteristics abundantly along his path. Once, I watched from a distance as he sat down next to a homeless person and spoke to her at length. He acknowledged her as a fellow human worthy of connection, then emptied his pockets and gave her everything he had. As he walked away she caught up to him, searched deep in her sack, and gave him what must have been her most valued possession: a small unopened plastic container of hand cream. That gift became one of Ali's dearest treasures. Now it's one of ours.

But now, because of a medical error, I'd lost him in the blink of an eye. Whatever I'd learned about happiness was going to be put to the test. I thought that if I could save myself and my family from the deepest depths of depression, I could count it as a great success.

But we did much better than that.

When Ali left our world so suddenly, his mother and I, as well as our daughter, felt profound grief. The pain of missing him still lingers, of course, and we regularly shed tears that he's no longer available for a hug, a chat, or a video game. The pain we feel drives us to honor his memory and wish him well. Remarkably, though, we've been able to maintain a steady state of peace—even happiness. We have sad days, but we don't suffer. Our hearts are content, even joyful.

Simply put, our happiness model came through for us. Even during the moments of our most intense grief over Ali's passing, we were never angry or resentful of life. We didn't feel cheated or depressed. We went through the most difficult event imaginable just as Ali would: in peace.

At Ali's memorial, hundreds of people filled our home to pay their respects while a huge overflow crowd waited outside in the 110-degree heat of Dubai's summer. They just would not leave. It was an exceptional memorial, in all ways built around the happiness that Ali had radiated throughout his life. People showed up in tears but quickly blended into the positive energy of the event. They wept in our arms, but when we talked, and when they came to understand our view of these events, which was informed by our happiness model, they stopped weeping. They walked around the house admiring the hundreds of photographs of Ali (always with a big smile) on every wall. They tried some of his favorite snacks set out on tables, or picked up an item of his as a souvenir, and remembered all the happy memories he'd given them.

There was so much love and positivity in the air, countless hugs and smiles, that by the end of the day, if you didn't know the circumstances, you might have thought this was just a happy gathering of friends—a wedding maybe, or a graduation. Even in these distressing circumstances, Ali's positive energy filled our home.

In the days after the memorial, I found myself preoccupied with the thought *What would Ali do in this situation?* All of us who knew Ali went to him regularly for advice, but he was no longer with us. I desperately wanted to ask him, "Ali, how do I handle losing you?" even though I knew his answer. He would just say, *"Khalas ya papa"*—It's over, Dad—"I've already died. There is nothing you can do to change that, so make the best out of it." In moments of quiet, I could hear no other voice in my head but Ali's repeating these sentences over and over.

And so, seventeen days after his death, I began to write. I decided to follow Ali's advice and do something positive, to try to share our model of happiness with all of those who are needlessly suffering around the world. Four and a half months later I raised my head. I had a first draft.

I'm not a sage or a monk hiding away in a monastery. I go to work, fight in meetings, make mistakes—big mistakes that have hurt those I love, and for that I feel sorrow. In fact, I'm not even always happy. But I found a model that works—a model that had seen us through our grief, the model that Ali's life helped generate through his example. This is what I want to offer you in this book.

My hope is that by sharing Ali's message—his peaceful way of living—I may be able to honor his memory and continue his legacy. I tried to imagine the positive impact spreading this message could create, and I wondered if maybe it is not for nothing that I have a high-profile job with global reach. So I took on an ambitious mission: to help ten million people become happier, a movement (#10million happy) that I ask you to join so that together we can create a small-scale global pandemic of Ali-style joy.

Ali's death was a blow I never could have expected, but when I look back, I feel that he somehow knew. Two days before his unexpected departure, he sat us all down as a wise grandfather would gather his children and said he had something important to share. He said he understood that it might seem odd for him to offer advice to his parents but that he felt compelled to do it. Usually Ali spoke very little, but now he took his time and spent most of it telling Nibal, Aya, and me what he loved most about us. He thanked us kindly for what we had contributed to his life. His words warmed our hearts, and then he asked each of us to do some specific things.

His request to me was "Papa, you should never stop working. Keep making a difference and rely on your heart more often. Your work here is not done." He then paused for a few seconds, sat back in his chair—as if to say *But now my work here is done*—and said, "That's it. I have nothing more to say."

This book is my attempt to fulfill the task assigned to me by my happiness idol. For as long as I live, I will make global happiness my personal mission, my moonshot for Ali.

Part One

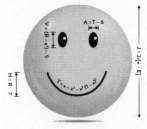

Happiness in the modern world is surrounded by myths. Much of our understanding of what happiness is and where to find it is distorted.

When you know what you're looking for, the quest becomes easy. It may take time to unlearn old habits, but as long as you stick to the path, you'll get there.

Chapter One

Setting Up the Equation

It doesn't matter if you're rich or poor, tall or short, male or female, young or old. It doesn't matter where you come from, what you do for a living, what language you speak, or what tragedies you've endured. Wherever you are, whoever you are, **you want to be happy.** It's a human desire about as basic as the drive to take the next breath.

Happiness is that glorious feeling when everything seems right, when all of life's twists and turns and jagged edges seem to fit together perfectly. In those often all-too-brief flickers of genuine happiness, every thought in your head is agreeable, and you wouldn't mind if time stood still and the present moment extended forever.

Whatever we choose to do in life is ultimately an attempt to find this feeling and make it last. Some people look for it in romance, while others seek it in wealth or fame, and still others through some form of accomplishment. Yet we all know of people who are deeply loved, achieve great things, travel the world, snap up all the toys money can buy, indulge in every luxury, and still long for the elusive goal of satisfaction, contentment, and peace—also known as happiness.

Why should something so basic be so hard to find?

The truth is, it isn't. **We're just looking for it in the wrong places.**

We think of it as a destination to reach, when in fact it's where we all began.

Have you ever searched for your keys only to realize they were in your pocket all along? Remember how you removed everything from your desk, searched beneath the couch, and got more and more frustrated the longer they went missing? We do the same thing when we struggle to find happiness "out there," when, in fact, happiness is right where it's always been: inside us, a basic design feature of our species.

Our Default State

Look at your computer, smartphone, or other gadgets. They all come with preferences preset by the designers and programmers. There's a certain level of screen brightness, say, or a localized user interface language. A device fresh from the factory, set up the way its creators think best, is said to be in its "default state."

For human beings, simply put, the default state is happiness.

If you don't believe me, spend a little time with a human fresh from the factory, an infant or toddler. Obviously, there's a lot of crying and fussing associated with the start-up phase of little humans, but the fact is, as long as their most basic needs are met—no immediate hunger, no immediate fear, no scary isolation, no physical pain or enduring sleeplessness—they live in the moment, perfectly happy. Even in distressed parts of the world, you can see children with dirty faces using little pebbles as toys or holding a cracked plastic plate as the steering wheel of an imaginary sports car. They may live in a hovel, but as long as they have food and a modicum of safety, you'll see them run around hooting with joy. Even in news coverage of refugee camps, where thousands have been displaced by war or natural disaster, the adults in front of the camera will appear grim, but in the background you'll still hear the sounds of kids laughing as they play soccer with a knot of rags for a ball.

But it's not only kids. This default state applies to you too.

Look back into your own experience. Summon up a time when nothing annoyed you, nothing worried you, nothing upset you. You were happy, calm, and relaxed. The point is, you didn't need a *reason* to be happy. You didn't need your team to win the World Cup. You didn't need a big promotion or a hot date or a yacht with a helicopter pad. All you needed was no reason to be *unhappy*. Which is another way of saying:

Remember !

➤ **Happiness is the absence of unhappiness.**

It's our resting state when nothing clouds the picture or causes interference.

Remember !

➤ **Happiness is *your* default state.**

When you use a programmed device, you sometimes change its default settings without meaning to, sometimes so much so that certain functions become more difficult to use. You install an app that frequently connects to the Internet, and your battery life decreases. You download malware, and everything starts to go haywire. The same thing happens with the human default for happiness. Parental or societal pressure, belief systems, and unwarranted expectations come along and overwrite some of the original programming. The "you" who started out happily cooing in your crib, playing with your toes, gets caught up in a flurry of misconceptions and illusions. Happiness be-

comes a mysterious goal you seek but can't quite grasp, rather than something simply there for you each morning when you open your eyes.

If we were to picture it, the times when you're unhappy are like being buried under a pile of rocks made up of illusions, social pressures, and false beliefs. To reach happiness, you need to remove those rocks one by one, starting with some of your most fundamental beliefs.

As every person who's ever called Tech Support knows, sometimes the first step to bringing a device back to proper functioning is to restore the factory settings. But unlike our gadgets, we humans don't have a reset button. Instead, we have the ability to unlearn and reverse the effects of what went wrong along our path.

How did we ever get the idea that we have to look for happiness outside us, to strive for it, reach it, achieve it, or even earn it? How did we get things so terribly wrong that we've accepted that happiness touches our lives only briefly? How did we let go of our birthright to be happy?

The answer may surprise you: *Perhaps that's what we've* **always been trained to do**.

Solve for Happy

You may have received sound advice like the kind my mother gave me, that I should study and work hard, save and be willing to defer certain forms of gratification to achieve certain goals. Her advice surely was a major contributor to my success. But I misunderstood. I thought she meant that I needed to defer happiness along the way. Or that happiness would be the result once I had achieved success.

Some of the happiest communities in the world are actually in the poorer countries of Latin America, where people do not seem to think much at all about financial security or what we consider success. They work each day to earn what they need. But beyond that, they prioritize their happiness and spend time with their family and friends.

I don't mean to romanticize a life that appears quaint and colorful but still falls below the poverty line. But we can learn from a mind-set that weaves happiness into each day, regardless of economic conditions.

I have nothing against material success. Human advancement has always been driven by innate curiosity, but also by the perfectly reasonable desire to store up enough resources to survive winter or a drought or a bad harvest. Thousands of years ago the more territory your family or tribe controlled and the better your skills at hunting and gathering, the better were your chances of survival. Thus the idea of sitting idle under the mango tree lost ground to the idea of innovating and hustling a bit, expanding one's territory, and building up a surplus, just in case.

As civilization developed, more territory and more wealth usually meant better living conditions and the prospect of a longer life. Eventually, capitalism came along, reinforced by the Protestant ethic, which made prosperity a sign of God's favor. Individual effort and individual responsibility allowed the rise of what we now call income inequality, which increased the incentive to work even harder, if only to avoid being outpaced and crowded out by others. And once you'd risen, you certainly didn't want to fall back. Because as the competition increased, the traditional supports that had provided security through the family or village eroded.

The era just before our own saw the Great Depression and two world wars in quick succession, during which even those at the top of the income ladder had to worry about the basics. As a result, hardship shaped the priorities of an entire generation, underscoring the idea that what mattered most in life was to never endure such hardships again. *The "insurance policy" most widely adopted and passed along was called "success."*

Increasingly, as the twentieth century gave way to the twenty-first, the middle class raised their children to believe that the only logical course was to spend years in educational institutions to gain skills to be deployed in a lifetime of hard work in the hope of attaining security. We learned to make this path our priority, even if it made us unhappy, counting on the promise that when we finally achieved what society defined as success, then, at long last, we'd be happy.

Now, just ask yourself this question: How often do you actually see that happen? And instead, how often do you see a successful banker or business executive who's swimming in money but seems to be miserable? How often do you hear about cases of suicide of those who seemingly "have it all"? Why do you think this happens? Because the basic premise is flawed: success, wealth, power, and fame don't lead to happiness. As a matter of fact:

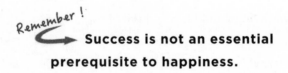

Remember!

➤ **Success is not an essential prerequisite to happiness.**

Ed Diener and Richard Easterlin's work on the correlation of subjective well-being and income suggests that, in the United States, subjective well-being increases proportionately to income—but only up to a point. Yes, it feels lousy to have to work two jobs to be able to afford a tiny apartment and a beat-up Honda while paying off student loans. But once your income reaches the average annual income per capita, which in the United States today is about $70,000, subjective well-being plateaus. It's true that earning less can dampen your sense of well-being, but earning *more* is not necessarily going to make you any happier.[1] Which suggests that all the expensive things advertisers say are the keys to happiness—a better cell phone, a flashy car, a huge house, a status-worthy wardrobe—really aren't so important.

Not only are wealth, power, and lots of toys not prerequisites for happiness; if anything, the chain of cause and effect actually works the other way. Andrew Oswald, Eugenio Proto, and Daniel Sgroi from the University of Warwick found that being happy made people roughly 12 percent more productive and, accordingly, more likely to get ahead.[2] And so:

Remember !

While success doesn't lead to happiness, happiness does contribute to success.

And yet we continue to chase success as our primary goal. One of the earliest psychologists to focus attention on happy individuals and their psychological trajectory was Abraham Maslow. Back in 1933, he summed up our pursuit of success in one profound sentence: "The story of the human race is the story of men and women selling themselves short."

While a reasonable level of success is common in our society, those who achieve the highest levels of success often have one thing in common, one thing that differentiates them from the pack. They all, almost compulsively, love what they do. Many successful athletes, musicians, and entrepreneurs have achieved their success because they love what they do so much they become experts at it just because the activity itself makes them happy. As Malcolm Gladwell puts it in *Outliers*, if you spend ten thousand hours doing something, you become one of the best in the world at it.[3] And what's the easiest way to spend so many hours on one thing? Doing something that makes you happy! Wouldn't that be better than spending a lifetime trying to reach success in hopes that it will eventually lead to happiness? At work, in our personal life, relationships or love life, *whatever it is that we do*, we should directly:

Remember! **Solve for Happy.**

What Is Happiness?

At my lowest point, back in 2001, I realized that I would never restore the happiness that was my birthright if I didn't at least know what I was looking for.

So, being an engineer, I set out to develop a simple process to collect the data I needed to determine what made me happy. First, though, I hesitated because the technique was so simple it seemed almost childish. But then it occurred to me: if our model for the default condition of human happiness is the infant or toddler, maybe "childish," or at least "childlike," is not such a bad thing.

I started by simply documenting every instance when I felt happy. I called it my **Happy List**. You might want to do the same thing. In fact, why not take a moment right now, pull out a pencil and a piece of paper, and jot down some of the things that make you happy. As assignments go, this one's not too tough. The list can be nothing more than a string of short, declarative sentences that get right to the point and complete the phrase

"I feel happy when _____."

Don't be shy. There's no reason to feel inhibited because no one ever has to see your list. You can include the obvious stuff, like scratching your dog under her chin or watching a beautiful sunset, and simple things like talking to your friends or eating scrambled eggs. There are no wrong answers. Write as many as you can think of.

When you're done, at least for the first pass, go back and highlight a few items that, if you were forced to set priorities, would be at the top of the list of things that make you happiest. Those will make for a valuable short list that will prove useful in our later discussions.

Here's some good news already: the very act of creating your Happy List makes for a very happy experience, so much so that, when you're finished, you should feel energetic and refreshed. I work on my list at least once a week, adding new things. Not only does it put a smile on my face, but it helps me cultivate something that psychologists say contributes to happiness over the long haul: an attitude of gratitude, which happens when you acknowledge the truth about our modern lives and the fact that there is plenty to be happy about after all.

So go ahead and enjoy. I'll go make a cup of coffee and wait for you. *(By the way, I feel happy when I enjoy a quiet cup of coffee!)*

The Happiness Equation

My hunch is that your list consisted almost entirely of ordinary moments in life—a smile on your child's face, the smell of warm coffee first thing in the morning, the kinds of things that happen every day.

So what's the problem? If the triggers for happy moments are so ordinary and so accessible, why does "finding" happiness remain such a big challenge for so many people? And why, when we "find" it, does it so easily slip away?

When engineers are presented with a set of raw data, the first thing we do is to plot it and attempt to find a trend line. So let's apply this to your Happy List and find the common pattern among the different instances of happiness on it. Can you see the trend?

The moments that make you happy may be very different from the moments that make me happy, but most lists will converge around this general proposition: Happiness happens when life seems to be going your way. You feel happy when *life behaves the way you want it to.*

Not surprisingly, the opposite is also true: Unhappiness happens when your reality does not match your hopes and expectations. When you expect sunshine on your wedding day, an *unexpected* rain represents a cosmic betrayal. Your unhappiness at that betrayal might linger forever, waiting to be relived anytime you feel blue or hostile toward your spouse. "I should have known! It *rained* on our wedding day!"

The simplest way for an engineer to express this definition of happiness is in an equation—the **Happiness Equation**.

Which means that if you perceive the events as equal to or greater than your expectations, you're happy—or at least not unhappy.

But here's the tricky bit: it's not the event that make us unhappy; *it's the way we think about it* that does.

Happiness in a Thought

There's a simple test I use to reaffirm this concept. Call it the **Blank Brain Test**. It's very simple. Recall a time when you felt unhappy, for example, *I was unhappy when a friend was rude to me.* Take your time and dwell on the thought, turning it over in your head and causing yourself as much unhappiness as you can. Let it linger in the same way we often do when we let thoughts like that ruin our day.

Please take a minute to find one such thought—and please accept my apology that I'm asking you to think about something that upsets you. Now apply the Blank Brain Test: Without changing anything in the real world, remove the thought—even if for just an instant. How can you do that? Engage your brain in another thought (read a few lines of text like you're about to do here) or blast some music and sing along. Or try the Ironic Process Theory, in which you end up making yourself think about something by trying not to think about it." Keep telling yourself, *Don't think about ice cream. Don't think about ice cream. . . .* until you find yourself thinking of nothing but ice cream.

How do you feel now? For the brief moment that you stopped thinking about your friend's rude behavior, were you upset? I thought not. Although nothing changed but your thought, there was a change in how you felt. Your friend still was rude, but you didn't feel as bad anymore. Do you recognize what this means? Once the thought goes, the suffering disappears!

When a rude person offends you, he can't really make you unhappy, unless you turn the event into a thought, then allow it to linger in your brain, and then allow it to distress you.

Remember!
→ **It's the thought, not the actual event, that's making you unhappy.**

But thoughts are not always an accurate representation of the actual events. So a slight change in the way we think can have a drastic impact on our happiness. I know this because one of the happiest moments in my life was when my beautiful, classic Saab *got totaled* in a crash.

I loved that car. It was a 900 Turbo in British racing green with a beige soft top, and one day Nibal took it out and ended up in a head-on collision with a truck. My toy was gone, but I was deliriously happy because the airbags, seat belts, and all the other safety features Saab was known for had deployed exactly as planned, and Nibal walked out of the crash without a scratch. I lost my car, but so what? My beloved wife was spared!

Now consider this: if Nibal had parked the car somewhere and then it was smashed, I would have been devastated. The results would have been the same—wrecked car and safe Nibal—but my experience of it would have been very different. The event itself was irrelevant. It was the way I looked at it that mattered.

So here's the $50 million question: If events remain as they are, but changing the way we think about them changes our experience of them, could we become happy simply by changing our thoughts?

Of course! This is what happens all the time already.

When a rude person apologizes, the apology doesn't erase the event, but it does make you feel better, simply because the gesture changes the way you think about what happened. It brings the emotional world inside you and the world of events outside you into better alignment and balances out your Happiness Equation. You start to agree with the world. The way life is becomes more the way you want it to be, so you feel happy again—or at least no longer unhappy.

The same turnaround happens when you find out that the rude person didn't mean what he said or that you misunderstood what he meant. Not a syllable of what was said has changed, but the way you think about it does, balancing the equation and leaving no reason to be unhappy.

There is ample evidence that we can actually manage our thoughts. We do that whenever asked to complete a specific assignment (such as what you are now doing by instructing your brain to read these lines of text). We tell our brain exactly what to do and it complies. Fully!

Pain versus Suffering

Just as our Happy List consists mostly of ordinary stuff, there are plenty of moments in ordinary, everyday life that are not to our liking. Even babies, our model for the happiness default, have plenty of things that can make them cranky: wet diapers, being left alone too long, being hungry, not getting enough sleep. Those moments of discomfort may be short-lived, but they serve a crucial, practical purpose. The discomfort of a wet diaper prompts the baby to cry, which prompts the mother or father or babysitter to change the diaper, which means that the problem gets solved before it causes a rash. As soon as the immediate discomfort goes away, the baby goes back to being happy.

In a similar fashion, most of the everyday discomforts of adult life are not only transient but also useful. The pangs of hunger prompt you to eat. The crankiness of inadequate sleep pushes you to get to bed.

The prick of a thorn makes you pull back your finger, and the pain of a sprained ankle prompts you to give it a rest so it can heal. Even serious physical pain exists as an important form of messaging between our nervous system and our environment. Without pain to help us navigate dangers, we would inadvertently do all sorts of things to hurt ourselves, and we'd never have survived.

Remember !

➤ **As much as we hate it, pain and the discomforts of life are useful!**

But as it is, we hurt—we heal. You burn your finger, you put some ice on it, you're good to go. Once the tissue starts to repair itself and the inflammation or irritation goes away, the pain has served its purpose. The brain no longer feels the need to protect the injured area, so it suppresses the signals, and good-bye pain. Which is why, barring a serious injury or a chronic condition, physical pain is generally not an impediment to happiness.

It may be less obvious, but everyday emotional pain is similar in that it also serves a survival function. Being left alone too long could be dangerous for a baby, so extended solitude becomes frightening to her and she cries to summon the caretaker. As adults, the painful feeling of isolation, also known as loneliness, signals that we may need to change our ways, to reach out more and try harder to engage. Painful feelings of anxiety can prompt us to seriously prepare for upcoming exams or presentations. Feelings of guilt or shame cause us to apologize and make amends, thereby restoring important social bonds.

When you experience emotional discomfort, you feel a little bruised for a few minutes, hours, or days, depending on the intensity of the experience. But once you stop thinking about it, the feeling of hurt goes away. Once time passes and memory fades, you can acknowledge and accept what you've experienced, extract whatever lesson you can from it, and move on. **Once the pain is no longer needed, it naturally fades away.**

But that's not the case with suffering.

When we let it, emotional pain, even the most trivial kind, has the capacity to linger or resurface again and again, while our imaginations endlessly replay the reason for the pain. When we *choose* to let that happen, that's when we overwrite our default for happiness and reset the preference for *needless suffering.*

The vividness of imagination also allows us to magnify the suffering, if we choose to, by adding our own *simulated* pain: "I'm an idiot for hurting my friend. I'm not good for anything. I deserve to be punished and suffer." The incremental layer of internal dialogue only leads to deeper and longer suffering by brooding over the story until it makes us miserable. But make no mistake, the misery we feel then is not the product of the world around us—the event is already over while we continue to suffer. It's the work of our own brains. In that sense:

Remember! **We let our suffering linger as a form of self-generated pain.**

All the thinking in the world, until converted into action, has no impact on the reality of our lives. It does not change the events in any way. The only impact it has is inside us, in the form of needless suffering and sadness. Anticipating awful things in the future or ruminating about awful moments from the past is not the useful, instructive, and unavoidable experience of everyday pain. This prolonged *extension* of pain is a serious bug in our system because:

Remember !

➤ **Suffering offers no benefit whatsoever. None!**

The interesting thing is, just as we have the ability to engage in our suffering at will, we also have the ability to debug our pain systems if we put our minds to it. But we don't always make that choice.

Imagine that you need a root canal and the dentist offers you either (a) the standard procedure with a few days of recovery or (b) a root canal with additional bonus days of extensive excruciating pain. Why on earth would you ever choose (b)?

Sad to say, each and every day, millions of people do just that: they effectively go for the root canal with extras. It all begins when you accept the thought passing through your head as absolute truth. The longer you hold on to this thought, the more you prolong the pain.

The day my wonderful son left, everything went dark. I felt I had earned the right to suffer for the rest of my life, that I was given no choice but to close my door and decay. I was, in reality, given two choices: (a) I could choose to suffer for the rest of my life and it would *not* bring Ali back, or (b) I could choose to feel the pain but stop the miserable thoughts, do all that I could to honor his memory, and it would still *not* bring Ali back—though it would make the world just a little bit easier to endure. Two choices. Which would you choose?

I chose (b).

Please don't get me wrong. I miss Ali every minute of every day. I miss his smile and comforting hug at the times when I need them most. This pain is *very* real, and I expect it to last. But I don't resist it. I don't have incessant suffering thoughts in my brain to magnify it. I don't curse life and act like a victim. I don't feel cheated. I don't feel hatred or anger toward the hospital or the doctor, and I don't blame myself for driving him there. Such thoughts would serve no purpose. **I choose not to suffer.** It helps me put life in perspective and move positively forward, sending Ali my loving wishes and keeping a happy memory of him alive.

Would you make that choice in the face of tough times? Assuming you could and that it's possible, **would you make the choice to stop your own suffering?** I realize that you might have endured unbearable hardship in your life, the pain of loss, illness, or lack. But please don't let those thoughts convince you that you're *supposed* to suffer, that you don't deserve to be happy.

Remember!

→ **Happiness starts with a conscious choice.**

Life doesn't play tricks; it's just hard sometimes. But even then we're always given two choices: either do the best we can, take the pain, and drop the suffering, or suffer. Either way, life will still be hard.

Keep that in mind. **You know *what* to do.** Now I'll show you *how* to do it.

Chapter Two

6-7-5

A thought can take its thinker through years of suffering. Seeds of thought grow and grow until they become angry monsters. And yet we believe in our thoughts and let them take hold. Happiness depends entirely on how we control every thought.

But contrary to common belief, we don't just experience two moods, happiness and sadness. Depending on the kinds of thoughts we entertain, we may fall into a wider spectrum of states:

- Allow your thoughts to be affected by illusions and you'll be stuck in the *state of confusion*.
- Think negative thoughts and you'll end up in the *state of suffering* (unhappiness).
- Suspend your thoughts by having fun and you'll find yourself in the *state of escape*.

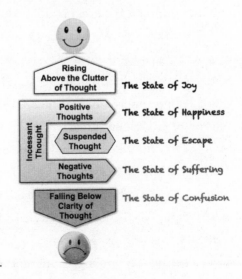

- Think positive thoughts and agree with the events of life and you'll reach the *state of happiness.*
- Rise above the clutter of thought, grasp life for what it truly is, and you'll perpetually live in a *state of joy.*

Understanding the difference between those states and the reasons you end up in one or the other will help you build a solid happiness model—one that will lead you to happiness every time you apply it. Let's investigate each of those states in detail, starting from the bottom and making our way up to the state of joy.

The State of Confusion

Do you sometimes feel that sadness is engulfing you as if you're stuck in quicksand? Do you sometimes feel that you're unable to shake away the fog that surrounds you, that blurs your vision and clutters your judgment? When you feel that life is against you and that you deserve to be miserable, you're in the state of confusion.

Our confusion is caused by illusions that we all learned to accept in early childhood. We learned to navigate the world believing the illusions are real. When you allow those illusions to inform your interpretation of the world around you, your judgment will lack objectivity, your attempts to solve for happiness will always yield incorrect results, and the resulting confusion will lead to deep suffering. Why do we learn to live with those illusions in the first place?

Imagine you were asked to take a leisurely drive around an empty racetrack. You, and most drivers, would probably cope well without understanding the basics of car mechanics or the G-forces affecting the turns. It's when things go wrong, or become challenging, that uninformed behavior is no longer sufficient. If the track got busy with fast drivers and the only way out was to race to the finish line, you would really need to understand how the car works at a fundamental level if you were to have any shot at making it off that track alive.

One example to help illustrate why things go wrong is the Illusion of Time. Most of us are constantly stressed by time's illusive nature. We run out of it, waste it, and feel it ticking faster every day, eating away at our stressful lives while unable to slow it down or stop it. The relentless pace overwhelms us. It truly feels like a fast track full of crazy drivers.

When we're caught up in an illusion, there's no point trying to solve the Happiness Equation. Life becomes so confusing that we don't even bother. We start to accept that we're supposed to be unhappy. Then our suffering lasts longer and runs deeper.

The State of Suffering

When a sad thought takes hold, we suffer. Then we let it linger. Why do we let thoughts prolong our pain when all we really want is to be happy? Why do we allow ourselves to worry about a test result when worrying will have no impact on the final outcome? Why do we obsessively recall an incident from the past, tormenting ourselves with regret, when our suffering can't affect what's already happened? Why do we let our thoughts deprive us of our childlike default state—being happy?

Keeping our negative thoughts alive, it seems, is just part of the original design of our human brain. The endless cycles of incessant thoughts are there to serve our most basic instinct: **survival**.

In the hostile environments our ancestors inhabited, they needed fight-or-flight responses to survive. The basic rules were these: It's safer to mark something as a threat when it isn't than to mark something as safe when it's a threat. And it's best to do that fast. As a result, their brains handled the information the real world presented to them in a way that was sufficient for survival, though it was not an accurate reflection of the truth.

Humanity's original survival programming lingers today. When we assess an event, our brains tend to err on the side of caution. We tend to consider the worst-case scenario so that we prepare for it, and we tend to morph the truth so that our limited brainpower can process it swiftly and efficiently. That's all well and good until you realize how often this leads to unhappiness.

Though some events do fail to live up to our expectations, we often give excessive attention to some that don't deserve it. Most events, when seen for what they really are, are absolutely consistent with how we *should* expect life to behave. For those, nothing, absolutely nothing, is truly *wrong*, perhaps other than the way we think about them.

We keep those events alive and painful, and we get stuck, suffering, believing that our *imaginary* perceptions missed our expectations.

The original design of the human brain included features that ensured the survival of our species. Those same features have turned into *blind spots* that delude the way our brain operates today. Distracted, our brain rarely ever tells us the truth, and that constantly ruins our Happiness Equation.

The point in our conversation when I reveal the blind spots and show you how to fix them will be a lot of *fun*.

The State of Escape

Speaking of fun, the modern world's favorite pastime, here is a big misconception that derails us away from the happiness we seek: **Often what feels like happiness actually isn't!**

We can miss the distinction between *happiness* and *fun*. We swap out true happiness for weapons of mass distraction: partying, drinking, eating, excessive shopping, or compulsive sex.

Biologically speaking, feeling good plays an important role as part of our survival machine. Our brains use it to drive survival behaviors that do not relate to immediate threats. To achieve that, our brains flood our bodies with serotonin, oxytocin, and other feel-good chemicals during acts they want to encourage us to do more often. Reproduction, for example, is vital for our species, but living without children represents no immediate danger to prospective parents. Without the pleasure associated with sex, such an important survival function would have been ignored. Mating gives us pleasure—and that urges our species to reproduce and propagate.

And so fun is useful, but some people seek it in desperation, *to escape*, because they're afraid of their difficult thoughts. In that sense, the fun they chase is like a *painkiller*, to blunt the suffering. Fun is an effective painkiller because it mimics happiness by switching off the incessant thinking that overwhelms our brains—for a while.

Remember!

⟶ **With no thoughts, we return to our default, childlike, state: happiness!**

As soon as the immediate pleasure fades, however, the negative thoughts rush back in and reestablish the suffering. So we keep coming back for more.

Just as with painkillers, when the effect of one fades, you pop another one until, eventually, taking regular-strength painkillers fails to numb the pain. That's when we try to inject more extreme pleasures into our life: extreme sports, wilder parties, and all forms of excessive indulgences. The more intense the high, the quicker the effects will fade and the deeper we will plunge back into suffering. When that cycle becomes too much to bear, some resort to desperate measures and chemically numb their brain using real drugs or alcohol in a final attempt to find silence inside their head.

By resorting to fun as an escape, we leave our Happiness Equation unresolved and ignore the core issues that make us unhappy. Fun, then, despite its short jolts of elation, truly becomes an inhibitor to genuine happiness.

But fun is not all bad. As a matter of fact, fun in itself is not bad at all.

A wise use for fun is as an **emergency off switch** to allow for momentary intervals of peace so that you can get the voice in your head to chill, meanwhile interjecting some reason into the endless stream of chatter. Whenever you feel the thoughts in your head getting negative, enjoy a healthy pleasure—say a workout, music, or a massage—and that will always flick off the switch.

An even wiser use for fun is when you actively schedule regular doses of healthy pleasures, which I define as pleasures that do not lead to hurting others or hurting you eventually. Fun, then, can be less like a numbing painkiller and more like a *happiness supplement* that you take regularly to stay healthy. As a businessman I've learned that we can improve only on that which we measure. So set yourself a *fun quota*. I do! I aim for a daily target of music and a weekly quota of comedy, workouts, and other feel-good activities. With enough pleasure in your life, the extended peaceful intervals make it harder for your brain to hijack your day with its uninterrupted streams of chatter.

But always remember: Fun and pleasure in any form are only ever a temporary state of escape—a state of unawareness. So never stay there for too long. Zoom through as fast as you can on your path to genuine, enduring happiness.

The State of Happiness

Happiness is all in a thought—the *right* thought—one that aligns with reality and solves the Happiness Equation positively. Funny, but we won't directly discuss happiness in this book. We will discuss how to stop the suffering, which will restore your default state of happiness. When you see the truth of your unfolding life and compare it to realistic expectations of how life actually unfolds, you will remove the reasons to be unhappy and realize, more often than not, that everything's fine, and so you will feel happy. For each event of our passing life, we solve the Happiness Equation correctly when we bust the illusions and fix the blind spots. But to stay happy regardless of the twists and turns of life, we should aim to reach an even higher state.

The State of Joy

Those who reach joy are not only accepting of life as it actually is but are utterly immersed in it. They are like the artists and writers—and engineers—psychologist Mihaly Csikszentmihalyi writes about, who are so in harmony with the present moment that they enter a realm of timeless bliss he calls "flow"—only they flow with every tiny thing that life throws their way, whatever it is. They reach a state of uninterrupted happiness that I've come to call *joy*.[1]

I use the term *joy* loosely here because unfortunately the English language is not equipped with a term that describes this state accurately. Inner peace, stillness, calm—these are all close. Perhaps a mixture of all of them is the closest, but none of them alone captures the true meaning.

A friend of mine was born with no sense of smell. One day she asked me to describe to her what it was like to smell a rose. I struggled to find the words. What can you say? A rose smells like, well, a rose! The only way to appreciate the smell of a rose is to experience it. It's the same with joy. All I can do is to help you experience it once, and then you'll know what it is.

When navigating an unfamiliar neighborhood, you constantly assess your current location on the map to find your way around. Block by block, you compare the map to the way the neighborhood is laid out. This is similar to what you do when you solve your Happiness Equation, thinking through event by event as your life unfolds.

But when the path is familiar, and you're in sync with it, you don't need a map. All you need is to orient yourself to a few major landmarks and follow your nose effortlessly to your target without much thought at all.

And that's the case with joy. It emerges, first and foremost, from a deep understanding of the exact topology of life. It comes as a result of having analyzed the Happiness Equation from a 20,000-feet view and completely grasped that life, with its mighty wheels of motion, *always* behaves as it *always* has and *always* will. As a result, you set realistic expectations; then, even when life is harsh, it no longer takes you by surprise because you realistically expect a bit of harshness along the path.

Annoying bumps on the road also arise when navigating a familiar path—they're not pleasant, but they are predictable, so you pass them calmly with no stress. The long time it takes you at the checkout counter of the supermarket, you realize, is what you should expect, and so should you expect work to be demanding, your manager to be annoying, and the money to run out at the end of the month. It's just how things are—bumps on the road of life. No surprises.

If fun suspends your thoughts, and happiness arises when your brain agrees with the events of your life, then joy is when thoughts are no longer even needed because the analysis has ended, and the equation has permanently been solved.

My wonderful son, Ali, had a tattoo on his back that he lived by: **The gravity of the battle means nothing to those at peace.** That tattoo described him fully. With that belief he went through life like  a wise old sage. Nothing could disturb his uninterrupted calmness. He rose above thought, and there he found joy.

The biggest myth about joy is that it's reserved for monks who give up on the fast track of life. But that's not true. Joy can be woven into everything that you do—even in the most stressful of all lifestyles.

When I used to trade in the stock market, my first big loss took me by surprise. I spent days suffering, regretting my actions and blaming myself. But then I went on trading for years where I occasionally experienced much deeper losses than the first, but remained completely calm and composed. Once you know the true nature of the market and that occasional losses—"ripples," as I used to call them—are just part of how the game is played, you stop the localized suffering and focus on the bigger picture. While the life of a trader is rarely ever joyful, that ability to form a realistic expectation of the risk inherent in the market and to rise above the ripples when they occur is the skill you need to reach joy.

Remember !
➥ **True joy is to be in harmony with life exactly as it is.**

But how do you find joy?

You do so by navigating life just like you would navigate a familiar path. You find your guiding landmarks—**you find The Truth**.

A Model for Happiness

Every day of your life, new events will unfold. New expectations get set, and new Happiness Equations demand solutions. Most of us randomly move to a different state with each passing event. We've all made a few steps forward to happiness . . . before plummeting into confusion. We've all found a shortcut by having fun for a brief moment . . . before experiencing a patch of suffering.

You've had enough of that, haven't you? A state of uninterrupted joy is attainable when you solve directly for it. And so . . .

Remember !
➥ **You should never settle for anything less than joy.**

But reaching uninterrupted happiness is not as easy as spending a night out with friends, attending a yoga class, or buying a new car. There are illusions to bust, blind spots to fix, painkillers to reject, and, finally, there are truths to ponder and grasp.

It's time to start your happiness training. As an engineer I'll give it to you in shorthand—in nowhere near as colorful a tone as the happiness gurus of today speak. It's not rocket science. All you need to do is remember three numbers: 6-7-5.

Here's how this works. There are **six grand illusions** that keep you in confusion. When you use these illusions to try to make sense of life, nothing seems to compute. The suffering runs deep and lasts long.

Next, **seven blind spots** delude your judgment of the reality of life. The resulting distorted picture makes you unhappy.

Eliminate the six illusions, fix the seven blind spots—and stop trying to escape—and you'll reach happiness more often than not.

But if you want your happiness to last, you must hang on to **five ultimate truths**.

Put it all together and you have the Happiness Model:

Bust the **6** Grand Illusions

Fix the **7** Blind Spots

Hang on to the **5** Ultimate Truths

Your training starts tomorrow.

See you at

You
Are
Here

Grand
Illusions

Blind
Spots

Ultimate
Truth

Part Two

Grand Illusions

6 grand Illusions submerge us in confusion and hinder our ability to make sense of the world. Life becomes a struggle. Most attempts to solve the Happiness Equation fail because we use illusions as inputs, unable to see the world for what it is, and we wonder why life has to be so cruel. When we see through those illusions a weight is lifted, our vision clears, and happiness becomes a frequent visitor.

Chapter Three

That Little Voice in Your Head

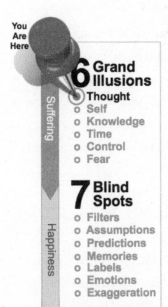

Listen.

Can you hear that voice?

The one right there inside your head?

Stop reading for a minute and try to enjoy a moment of silence. See how long the moment lasts before that voice pops up into your head to talk about all the things you need to do with your day, to remind you of the rude person you encountered around the corner, and to worry you that you're not going to get that promotion you've been waiting for.

Specific elements may differ, but the endless stream of chatter is something we all share. It worries us about what is yet to come; it belittles us; it disciplines us; it argues, fights, debates, criticizes, compares, and rarely ever stops to take a breath. Day after day we listen as it talks and talks.

While having a voice in your head is pretty normal, that doesn't make it a good thing. You shouldn't ignore the unhappiness, pain, and sorrow it's causing us. Should we?

Perhaps it's worthwhile to spend some time to try to understand more about that voice. Let's start with the basics: Who's talking? Is that voice *you* talking to *you*? Why would you need to talk to yourself if you're the one doing the talking?

The Voice Is Not You

If there is one thing that will change your life forever, it is recognizing that the voice talking to you is **not you!**

Think about that for a minute. It's so simple it doesn't even need proof. A vantage point is a prerequisite to perception; to observe something you need to be outside of it. We could not see planet Earth as long as we never left it. Only when astronauts sent us back pictures of Earth could we see it. You can't see your own eyes because they are the part of you that sees. The image of them reflected back to you in a mirror is just a reflection. It's not your own eyes.

If you can hear someone speaking on the radio, that someone isn't you. Similarly, for you to perceive a voice speaking in your head, you and the voice must be two separate entities.

Not convinced? Then consider this: What happens when, for a few seconds, you stop thinking? We all do this sometimes. Does that mean for those short moments you cease to exist? That you're no longer you? Who, then, is enjoying the silence? The answer is *you*. The real you. When you open your eyes in the morning for that brief moment before the stream of thought commences and you look at the alarm clock—who is looking? Who notices the sunshine outside before thought takes over and starts to narrate the day? The same person who has to listen to the nonstop chatter of that little voice in your head for the rest of the day. This concept will shortly become clearer, when we discuss who the voice is. But for now the truth is simple:

Very Important !

↳ **The little voice in your head is *not you*!**

Even if this point seems simple, it should revolutionize the way you approach your thoughts. Modern culture drastically overvalues logic and thought. We even go as far as to equate our own being with thought. René Descartes's famous assertion "I think, therefore I am" seems to find a lot of acceptance in brain-dominated Western culture—but is it true?

When you believe that you are your thoughts, you identify with them. In other words, if you have a thought that seems naughty, you might think that *you* are naughty. Get it? But naughty thoughts don't equal a naughty person. Naughty thoughts are simply presented to you for your consideration; that's what the brain does. What you do with those thoughts is up to you. **You don't have to obey.**

When you finally realize that you aren't your thoughts, you'll have seen through the most profound illusion of all: the Illusion of Thought. You are not your thoughts. Those thoughts exist to serve you.

What Descartes should have said is:

Remember !

↳ **I am, therefore I think.**

But if the voice isn't you, who is it? In cartoons it's depicted as an argument between a little devil on your left shoulder and a little angel on your right, each whispering their own agenda in your ears. In *A*

New Earth, Eckhart Tolle calls that voice "the Thinker"; some religions think of it as the devil making the case for his devious plans. Others have called it "the Whisperer" or "the Companion." The one thing all those names have in common is that they identify that voice as a separate entity, one that tries to talk you into doing things you wouldn't otherwise do without a bit of convincing.[1]

A friend of mine calls the voice in her head "Becky." When I asked her why, she said it's the name of the girl she least liked back in high school, the one who always pushed her to do things she didn't want to do.

Please feel free to call your voice what you want. The exact nature of it is irrelevant for the remainder of our conversation. What matters is that you recognize it exists, acknowledge that it isn't you, and understand how it behaves. I simply call it the brain—because that's what it is.

The Brain

Made up of more than 200 billion neurons with hundreds of trillions of connections among them, the brain is by far the most complex machine on the planet. If you count each neuron as a small computer, your brain would have thirty times more neurons than the number of computers and devices that make up the entire Internet.[2] It interfaces with your senses and controls your muscle functions, movements, actions, and reactions. It's capable of complex analyses, mathematical calculations, and logic, but also the negative kind of incessant chatter that holds you back from happiness. It's the most valuable instrument we've been given. Unfortunately, it didn't come with an operator's manual, and so very few of us truly learn how to optimize using it.

Imagine what a waste it would be if you were given the fastest sports car in the world and the only part of it you used was its audio system. Or imagine if you took it off-road, where it got stuck because this is absolutely not where it's built to go. Or, even worse, if you never received training as a racecar driver and drove like a maniac, so you hurt yourself and everyone around you.

We commit all three of these errors when using our brain. We use it for the wrong reasons; we don't utilize the best of its abilities; and we allow it to spin out of control with our thoughts—letting it ruin our lives and those of others'. We can do better than this, but first we need to understand why we use our brains the way we do.

To grasp why this complex machine talks so much, let's go back to the time when it didn't talk at all and observe a newborn child. Before we learn words, our brains are silent. We just lie there observing and interacting with the world. As we get older, we notice that our parents are busy using words to convey messages: *bottle, food, diaper, bath.* We are praised when we repeat those words, so we develop this skill of calling everything by its name, even if no one is around us to hear it. Words become our only method to understand and communicate knowledge. We start to narrate what we observe to help us make sense of things. As infants, we do that out loud; then, when it becomes socially awkward, we start moving the narration inside. From then on, it never stops.

In the 1930s, the Russian psychologist Lev Vygotsky observed that inner speech is accompanied by tiny muscular movements in the larynx. Based on this, he argued that inner speech developed through the internalization of out-loud speech. In the 1990s, neuroscientists confirmed his view; they used neuroimaging to demonstrate that areas of the brain such as the left inferior frontal gyrus, which are active when we speak out loud, are also active during inner speech. That voice inside your head truly is your brain talking, even though you're the only one who can hear it.

The Job Description

So we know who's talking, but why does it talk? Like other organs, your brain is there to perform a specific function. At its most basic, the brain's core task is to ensure the safety and survival of your body.

Some of that work takes place without your even knowing it. If your peripheral vision picks up a car speeding down the street toward you, your brain will order your legs to jump. Occasionally, when the threat warrants more than just a reflex, the brain triggers the release of adrenaline so you'll be ready for your fight-or-flight response. All those survival reactions are mechanical in nature; they take place without your having to make any conscious decisions. Very impressive!

Thought engages to add an extra layer of protection when the brain plans ahead to keep you away from possible danger. It assesses every cave, tree, rock, or wherever a tiger might be hiding. When you're looking out over a spectacular vista, your brain's first job isn't to kick back and enjoy it, but to consider any aspect of it that doesn't look right and might therefore pose a threat. It's also conditioned to consider long-term dangers so that we plan for winter, provide shelter to protect our young, and constantly analyze each of the countless things that might go wrong.

When external threats surrounded us in the earliest years of humanity, both forms of brain functionality were extremely vital for our survival as individuals and as a species. Fear kept you alive, and your brain was fully in charge. For reflexes, it didn't even consult with you, and still doesn't today. It just does what it is supposed to do. When it came to decisions that didn't address an immediate threat, however, your brain assessed the challenge more thoroughly, using two different approaches, one that is intuitive and quick and another that is slow and deliberate,[3] resulting in what seems like a dialogue.

> Hey, dude, remember that cool guy Tommy, who was shredded to pieces by the tiger? We don't want that to happen to us, do we?
>
> No, we don't.
>
> Good. See this tree here? It looks like the one the tiger jumped from behind to shred Tommy. Let's walk down by the riverbank instead. Would you like that?
>
> No, it's quicker through the jungle, and there's nothing to hunt by the river.
>
> Look, dude, Jessica is going to be back at the cave tonight, and I would much rather be there doing whatever it is that we do there than be shredded, so let's just go by the river today.
>
> Yeah . . . Jessica . . . Okay.

That kind of dialogue is the brain's attempt to arrive at the best possible decisions. Daniel Kahneman, winner of the Nobel Prize in Economics, explains this process brilliantly in his best-selling book *Thinking, Fast and Slow.* He discusses the dichotomy between two modes of thought: "System 1" is a fast, instinctive, and emotional mode of thought; "System 2" is a slower, more deliberative, and more logical mode. Often in his book he cites examples of mistakes or quick inaccurate judgments made by System 1 that are corrected by System 2. The presence of these two systems is what leads to your sometimes having two voices in your head. They're simply two modes of thought looking at the issue at hand from different perspectives and with varying skill sets, discussing it on the center stage of your head.

Smile, you're not crazy after all.

Who's the Boss?

Since the dawn of humanity, our brains have assumed full responsibility for our existence, and because survival was so much more tenuous in the beginning, we accepted our brains as our undisputed leaders. But is that still justified?

It's indisputable that your brain does some things very well, but it shouldn't be given the freedom to deliberate everything. When it manages reflexes and mechanical functions, it does so without thinking. This is true for all vital functions—thought is completely left out. The work of your lungs, glands, heart, liver, and other organs is mechanically run by your brain but isn't the result of conscious thought—you don't spend hours dwelling about them or even have the ability to command their functionality. If the brain were allowed to control them, it would make massive mistakes. It might, for example, in a moment of huge emotional pain, make the seemingly logical decision to end your life by switching off your heart. Luckily, this feature is excluded from our design because thought does *not* always produce the best results.

Remember !

The more something matters, the more incessant thought will be left out of it.

Have you ever noticed that?

Well, guess what: **Happiness really matters.** Why, then, do we let our thoughts sometimes burden us and deprive us of it? Accept your brain as the undisputed leader when it comes to mechanical operations, but when it comes to thought, *you* should be in full control. Your brain's job is to produce logic for you to consider. When the thoughts are presented, you should never lose sight of the question *Who is working for whom?*

Remember!

⮕ **You are the boss. You get to choose.**

This means you tell your brain what to do, not the other way around. Just as you are now instructing your brain to focus on the words on this page, you can always tell it what to focus on. **You just need to take charge** and act like the boss. Correct Descartes's statement all the way:

Remember!

⮕ **I am, therefore *my brain* thinks.**

Useful Thinking

To function well in the modern world you need to differentiate what's working for you from what's working against you. While it sometimes feels that all of our thoughts are an incessant stream of useless blabber, the reality is that our most useful thoughts are usually silent. There are three types of thought that our brains produce: insightful (used for problem solving), experiential (focused on the task at hand), and narrative (chatter). Those types are so distinctively different from each other that they occur in different parts of our brain. A study by researchers at MIT in 2009 revealed how insightful thinking works.[4] Brain signals of human participants were recorded while they solved verbal puzzles. Individuals were asked to say the answer out loud as soon as they reached a solution. The results showed that two regions of the brain, both on the right side, are involved in solving the puzzle. One brain region works in the background, but we become aware of the answer, as a thought, in the other region—up to eight seconds later.

More interesting, both regions where this kind of useful thinking happens are very different from the regions where incessant thought occurs. This was shown in a study conducted in 2007 at the University of Toronto, in which researchers monitored the brain functions of two groups of participants: a novice group whose incessant thought was active and a group who had attended an eight-week course that trained them to develop focused attention on the present.[5] The study found that the incessant thoughts of the first group lit up the midline regions of the brain, while the second group (who were adept at paying attention to the present moment) activated the right side of the brain, and regions different from those used for insightful thinking.

Now here is the good news: incessant thought, being just a simple brain function, offers strong evidence that our thoughts are in no way who we are—they don't define us. Once again, *you are not your thoughts*. Your brain produces thoughts, as a biological function, to serve you. And discovering that each of those types of thoughts happens in completely separate brain regions means that we can be trained to use one type more than the other.

We need a lot of attention to the present when we perform tasks, and we need problem solving. Those are very useful functions. What we don't really need is the narrative component of thought, the useless, endless chatter—the part that makes us feel a bit crazy and keeps us trapped in suffering.

The Suffering Cycle

When our ancestors recognized a threat in the hostile environments they inhabited, it triggered a physical fight-or-flight response. In the modern world, most of the events we encounter represent a threat only to our psychological well-being or to our ego. Frequently no proper survival mechanism can protect us from such threats. In the absence of a satisfactory response, our brains tend to bring the unresolved threat back over and over in a constant stream of incessant thought.

As per the Happiness Equation, the re-petitive loop of thinking of an event, com-paring it unfavorably to our expectations, leads to suffering. Our inability to take action triggers the recall of the thought over and over in an endless **Suffering Cycle**.

We can disrupt this Suffering Cycle by neutralizing the negativity at each of its nodes.

Taking the best possible action, regardless of the result, is an obvi-ous way to disrupt the cycle. Once an action is taken, our minds focus on the executional elements of what needs to be done, a different part of our brains is engaged, and our thoughts shift to monitor the result of the action rather than incessantly focusing on the same thought.

Another way is to stop the thought from turning into suffering. This can be achieved by fixing our blind spots to ensure that the events are seen for what they really are, not what our brain makes them appear to be. This is the topic of chapter 9.

But why let the cycle start in the first place? Wouldn't it be better if that little voice were quieter?

Managing the Voice

If you think about the degree of control you have over your heart and muscles, you'll notice that there is a difference between them. Your heart will always beat; it isn't within your control to stop it. It's an *autonomous* device. Your muscles, on the other hand, are partially within your control. Though reflexes force your muscles to act in ways you don't intend for them to act, you can command your arm to carry a weight when you so please. Even if it is heavy, you can push your muscles to perform a bit better. You have many such systems in your body. I call them *controllable* devices.

This is a fundamental difference.

Your brain belongs to the category of controllable devices because you have partial control over it. You can tell it what to think about, how to think, and even to stop thinking altogether. You just need to practice that control until you master it. It's doable. Isn't this incredible news?

Brain control may sound a bit like a theme for a science fiction movie, but you do it every day of your life. Focusing on homework or doing financial planning or discussing a specific topic with a friend are all examples of exercising control over your brain and telling it what to do. You can practice such control over the voice in your head as well.

What follows are four techniques to achieve that. Each builds on the one before it, so master them in order. They're simple but require discipline. Practice makes them easier until they become second nature. When you stop practicing them for a while, your brain will try to go back to its old habits and occasionally it'll succeed. Don't be alarmed. Just kindly and gently tell it, "I see what you're doing here. I know this is hard for you. If you play along for now, it'll be better for both of us eventually."

Observe the Dialogue

First, take your time to become very familiar with the beast you are taming. The best way to do that is to sit quietly and observe what is going on up there as often as you can. This technique is called "observing the dialogue."

Don't resist the thoughts that'll pop up. Instead, keep watching them as they roll on through. Observe a thought—then let it go and remind yourself that this thought isn't you. Thoughts come and go. They have no power over you unless you give them power.

When you master the technique of observing the dialogue it will feel as if you're watching *Seinfeld* (my favorite American sitcom from the 90s), a show about nothing. You follow attentively, laugh frequently, and still have no need to participate. You don't *judge* what is being said or interrupt to debate a specific bit of dialogue. Let your brain speak like the characters of a sitcom.

Now that you know the thoughts are not you, it's much easier to keep from getting bothered or annoyed. Observe each thought as it comes—then let it go. Do this on your daily commute, when you have to wait for your next appointment, or whenever you have a few minutes free. Make it your favorite pastime, your very own private sitcom, your "show about *nothing.*"

Here's the best part: as soon as you master the art of observing an idea and letting it go, your mind will quickly run out of topics to bring up. It can keep going only when you cling to an idea. You'll be surprised at how quickly your brain becomes domesticated. It'll slow down its wild, aggressive, incessant stream. Once you feel that, move on to the next technique.

Observe the Drama

No one's able to let go of every thought. Occasionally an idea will stick. You'll recognize the signs: you'll be fully absorbed in thought and less aware of the rest of the world around you. When you notice this happening, this is your chance to learn to observe the drama.

Start by acknowledging how you feel, the emotion triggered by the thought. Don't resist it. Let it be. You may want to dig deeper, not in an attempt to solve the problem but to try to understand it better. Ask yourself why you became angry or agitated. Which thought led you here?

For a long time I used to get annoyed by the sound of children crying or playing around me whenever I went to a café to enjoy some quiet time. They seemed to show up wherever I went. Believe it or not, even as I write this I'm in an almost empty café—except for some kids shouting at the table right behind me. In the past I would have been boiling with angry thoughts. *Are those parents going to do anything about it? Don't they have any sense of responsibility or respect for others?*

The more my thoughts lingered, the angrier I became, until one day I learned to observe the drama. Instead of focusing on the noisy kids, I learned to observe the thought that triggered my anger. Then I asked myself, *Why am I having these heated emotions? Why do I get this angry? Why do the screams of children annoy me where loud music doesn't?* (I'm a serious heavy metal fan. It doesn't get any noisier than that.)

And then it all became clear.

When I was a younger parent, my sunshine, Aya, was full of energy. (She still is.) Whenever we went out, she would be the one making noise. I remember how embarrassed and uncomfortable I felt. It hurt my ego to be the father who wasn't able to "control" his child. It made me feel guilty because I really didn't want to spoil anyone's quiet time.

Now I was the other character in my embarrassment, the figure whose peace I was ruining. Years later my brain still associated the screams of a young child with those feelings of embarrassment and guilt. Bingo!

Once I saw the *reasons* for my feelings, they became easy to navigate. Children don't bother me anymore. They scream and shout—and I stay calm. These days, those noises bring back memories of how incredibly talented Aya was as a child, and I smile. I remember how she used all of that energy to become the artist that she is and how she grew to travel the world even more than I did as a result of her restlessness. The same event that once brought me anger now triggers happiness. Reframing the thought reframes the emotion.

Now there's another family pulling their stroller to the table right next to mine. I swear I'm not making this up. There comes the noise, and here comes my smile. I miss you, little Aya.

Start observing the drama. The simple act of trying to trace the emotion to the thought that caused it gives you the breather you need in order to cool down. Focusing on connection uses the problem-solving side of your brain, and so it helps you stop the incessant chatter as it helps you to pinpoint that originating thought. When you clearly observe it, you'll realize that it's often not accurate, and certainly isn't worth the price you're paying to keep it simmering.

As you grow accustomed to this practice, you'll notice the repetitive patterns of your brain. You'll be able to read the tricks of your brain like an open book, and when it engages them you'll simply smile and say, "He-he, you're so silly, brain! Why don't you go bring me a better thought?"

Bring Me a Better Thought

Once a negative thought takes hold, it can become hard to dispatch. An untamed brain needs a thought to cling to. And often enough, removing a thought leaves behind a vacuum that gets quickly filled by a thought from the same mood spectrum—another negative thought. This is why when you are in a dark place it can seem like the whole world is going to collapse. You tend to be consumed by one negative thought after another. If only you could manage to break the cycle! Filling that vacuum with a happy thought ensures that there is no room for another negative one to come in.

This is when the fun begins.

It's simple. Look at the bolded word below. Take a few seconds to focus fully on it.

ELEPHANTS

If I may ask, what are you thinking about now? Could it possibly be an elephant? Regardless of what you were thinking about before, I can guarantee that your thought changed when you read the word *elephants.*

Remember !
→ **Your brain can be primed!**

Simple as this might seem, this is a powerful loophole in your brain's thought cycles. The effects of this secret back door are extremely predictable. Every single time your brain is tempted with a thought it will take the bait. It can't help itself! We can put this to very good use. You can prime your brain to focus on anything you want just by bringing it into consciousness.

With infinite choices, what should you tell your brain to think about? Yep, you got it:

Remember !
→ **Happy thoughts.**

If you could prime your brain with any thought you wanted to, why would you ever prime it with anything else?

Once, when Aya was around five, she was crying while I was deeply engaged trying to explain to her why she shouldn't cry about the issue that had upset her. In the cutest way she looked at me with tears in her eyes and said, "Papa, when I'm crying don't talk to me about the things that make me cry. If you want to make me happy, *just tickle me.*" Of course! This simple nugget of wisdom has stuck with me. We believe we need a solution for our unhappiness to go away, but often the reason we're unhappy isn't justified, and therefore there is no real solution for it, just as there wouldn't be for a false premise. So the easiest way to become happy is to *just be happy.* Remove the unhappy thought, replace it with a happy one, and let the rest take care of itself.

From now on, whenever a painful thought comes up, simply prime your brain to think about something else. Sometimes life deserves nothing more!

There is an important detail to keep in mind, though: deeper thoughts happen in the unconscious part of your brain. Unlike your conscious mind, which traffics in words, your unconscious mind developed long before you could use words, so visual images and sensations are its currency. This matters because there's no image that corresponds to the word *no*. Your unconscious brain can't process a negative. In your *conscious* mind you can simply negate a concept, as in "*no* suffering." But your unconscious mind would take that concept and think only of the word it understands—the very word you want to negate: *suffering*. Instead of negating a concept, you have to replace it with the opposite of that concept. As far as your unconscious mind goes, you can't think of *no suffering*; you can think only of *happiness*. Instead of trying to think about not being at a job you dislike, think about being in another job altogether. Instead of thinking about ending a relationship, think about the new one you would like to start. That's the way to shift your thinking into happy thoughts.

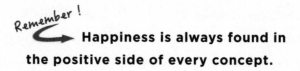

Remember !

Happiness is always found in the positive side of every concept.

The easiest way to have a full arsenal of happy thoughts is to use your Happy List (from chapter 1). A happy thought doesn't need to be related in any way to the dark topic that derailed you. Any happy thought on the list can short-circuit the stream of negativity in your brain by filling the vacuum. Once the negative stream of thoughts has been broken, you'll find it much easier to resume thinking with a positive outlook.

If you find this technique difficult at first, write down your Happy

List on an index card and carry it with you all the time. Or you know what works even better? Carry pictures of your happy thoughts on your phone so they're always ready for you.

For years I went everywhere with a folder of nineteen happy thoughts in it. Now I don't even need to pull it out anymore; the right picture automatically pops up in my brain to push the negative ones out. When my state is reset back to a positive mind-set, I start focusing on the challenge at hand, specifically the parts that are firmly within my control, and apply positive energy and useful thoughts to make things better.

A better way to utilize your Happy List is to use it proactively instead of defensively. Take your list out several times a day and focus right there. You might get so good at this that you'll never need to wait until a negative thought arises. The longer you keep your brain in the positive zone, the more difficult it will become for it to shift into negativity, and the more that useless part of your brain will diminish (if you don't use it, you fortunately lose it).

With practice, you can take this process a step further. You can learn to prime your brain with happy thoughts related to the topic it has been thinking negatively about. All you need to do is prearrange a set of questions that probe the positive side of any issue.

Take, for example, the thought "I hate my job." If you left it up to your brain, it would take that thought and drive it further into all the things that make you miserable at work. Instead, prime your brain with a question like "There must be something that I like about this job. What is it?"

At first, your uncooperative brain will continue on its original path and bring another negative thought: "I hate the way my boss tells me to do things." In response to that, calmly insist (as though you were talking to a six-year-old), "So what do I *like* about that place?" Only then will you get back something at least partially positive, such as "The receptionist is nice, but man this place is ugly." Keep at it and the positivity will flow. "I like the coffee shop downstairs. The commute to work is easy. The money isn't bad." Hang on to those thoughts. This is finally working. You can now see the full half of the glass.

Usually things aren't all bad. Train your brain to find the good and make it the focus of your thought. In the same way you made your Happy List, make a list of generic questions that can prime for positive, such as "What is good about this situation? What do I like about it?" Or you can boil it down to one simple question: "What's in the full half of the glass?"

Once you get the hang of it, you'll become a master at finding the good side of things. They're always there; you just haven't been searching for them. Your brain will learn that negative thoughts don't get it very far and that the only way to get your attention is to think positive thoughts. It will be tamed.

When you find it easy to reroute the conversation, you'll be ready to streamline the process even further. The next time you notice a negative thought in your brain, simply respond with *Go get me a happier thought.* That's really all you need to say. As usual, your brain will try to evade the task at first, but if you insist, it will comply, and from then on all you'll ever need to do is to repeat the statement *Go get me a happier thought* until you get one. Those who manage to do this get as close to "brain control" as we ever get.

Congratulations, **You,** not your brain,

are now the **BOSS!**

Shut the Duck Up

If you've been observing the dialogue for a while now you'll anticipate what I'm about to say. Don't you sometimes feel there is a little duck in your head? And that it keeps quacking away all the time? It rarely ever gives you a single moment of peace. The

duck always quacks. After learning how to make my brain think positively, I once listened to Pete Cohen, author of *Life DIY*, discuss how the constant quacks affected the performance of the top-tier athletes he trained, and I found myself thinking, *By now I know how to make the duck quack positively, but Pete is right. Sometimes I wish I could just shut the duck up!*

There are many well-known meditation techniques that can help you practice that peace. They often involve focusing your brain on something outside the scope of thought: the beauty of a rose, the flickering flame of a candle, or your own breath.

Meditation isn't a lifestyle, though. It's a practice that prepares you for a lifestyle. What good is the practice if you go back to the normal "head full of thoughts" lifestyle once the practice is over? The ultimate aim is to live in a state of greater awareness outside the meditation room, so it becomes your lifestyle throughout the day.

Another quirk in the way the brain works can help you achieve that. The brain is what's known in computer science as a serial processor, which means it can focus on only one thought at a time. Even though it may sometimes feel like you have a million thoughts in your head, what your brain is really doing is shuffling between them quickly.

Take a minute now to enjoy the following game. Try to think of any two things at the same instant. Try to think of the fun you had last weekend while keeping in mind the memory of that argument you had yesterday. Keep trying. Keep trying. Tricky, isn't it? Now try reading out loud while silently counting down from 643. You'll notice that when you read, the counting stops, and when you count, the reading stops. This is also the case with your internal dialogue. One thing at a time is all that this magnificent machine can do.

Remember!

→ **For the brain, multitasking is a myth!**

We can use this feature of our brain to our advantage. My recommended technique to shut up that quacking duck is to flood it with things that it can't think about, evaluate, or judge—things it can only observe. Here's how: Direct your attention outside yourself. Observe the light in the room, pay attention to whatever is on your desk, catch that smell of coffee percolating in the kitchen, notice the wood grain on the table, or listen to the distant sounds of cars in the street. Don't let anything go unobserved. Notice every tiny detail around you. This is what you used to do as a newborn child. Just observe.

Alternatively, you can borrow from meditation techniques and turn your attention within. Pay close attention to your body. Tune in to any sore muscles from yesterday's workout or back pain from sitting at your desk too long. Observe your breathing or feel the blood pumping through your body.

Take it all in: the infinite stimuli that your brain has been filtering out so it can free up the brain cycles it needs to obsess over its own thoughts. Choose the one thing it can process at a time to be something *other than thought*. Flood it with signals from the physical world so it stops living in its own little bubble. Each filter you remove gives your brain something to process and reduces its ability to engage in useless thoughts.

This time you're not priming your brain with a good thought—you're priming it with no thought at all. This is when silence sets in. Big peaceful smile!

Be warned, though. This can be a very uncomfortable zone for your brain. After all, it's used to being the boss, and your ability to turn it on and off will feel like a threat to its existence. It will fight back by sending more thoughts your way. The best response is to stay still and calmly observe the world inside and outside. Keep removing those filters again and again until the silence returns.

By using this technique I have learned how to simply pull the plug on my brain, even after years of being a left-brained tech executive. Sometimes I sit for hours on a long flight with a dumb smile on my face and just the ghosts of thoughts—or none at all—in my head. It's *heaven*. An on-demand off switch for all thought. I just firmly tell my brain, "I want you to shut up, now!," remove my sensory filters, and enjoy the world without any brain commentary.

Try it yourself. It's a joy like no other.

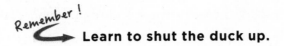

Remember! Learn to shut the duck up.

"The One"

In the 1999 sci-fi movie *The Matrix*, a simulation created by sentient machines is used to subdue the human population. Keanu Reeves plays Neo, known as "The One," and is chosen to free humanity. When he finally sees beyond the illusory images and thoughts planted in his brain by the Matrix, everything turns into ones and zeros right before his eyes. My programmer's mind saw this as his absolute clarity of vision, which led to his ability to take full control of his environment.

Nothing could harm Neo anymore. The blazing-fast movements of the Matrix "agents" became slow-motion, and he could effortlessly stop their punches and evade their speeding bullets.

This is the skill level you're moving toward as you start to see through the Illusion of Thought. So much of your happiness depends not on the conditions of the world around you but on the thoughts you create about them. When you learn to calmly observe the dialogue and the drama, you begin to see the ones and zeros. You can watch your thoughts, knowing that the only power they can gain over you is the power you grant them.

Like Neo, you will start to feel your thoughts flowing more slowly. You will observe each one and dodge its attack. More important, once you learn to order your brain to go out and get you better, more positive thoughts, you will reach that stage when you are in full command. You can tell your brain what to make of the world around it.

One thing that always struck me about *The Matrix* was how disinterested Neo's face became when he finally saw the world for what it is. While the agents were fully engaged both emotionally and physically as they attacked, Neo was unimpressed and unmoved by what the Matrix threw at him. He did what he had to do, knowing that the fight was already won. He was already at peace.

You too can be The One. You can stop the flight of your brain's speeding bullets and peacefully watch them in their raw format as they zoom past.

It may take you a while to get there. Until you do, you should have no other goal. This is the black belt of mind control. This is complete peace.

Please don't let the illusion fool you.

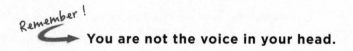

Remember! → **You are not the voice in your head.**

Who Are You?

When I speak about what it means to *not* be our thoughts, most people almost always smile with relief. They realize they don't have to listen to that duck anymore. But a moment later, a new confusion sets in. Their brain snaps back into hyper attack mode, posing a fundamental question: *If I'm not the voice in my head, then who am I?*

Good question. Think about it for a few minutes before you flip the page.

Chapter Four

Who Are You?

You Are Here

Suffering

6 Grand Illusions
- o Thought
- o Self
- o Knowledge
- o Time
- o Control
- o Fear

Happiness

7 Blind Spots
- o Filters
- o Assumptions
- o Predictions
- o Memories
- o Labels
- o Emotions
- o Exaggeration

5 Ultimate Truths
- o Now
- o Change
- o Love
- o Death
- o Design

Joy

This, no doubt, is one of the most pivotal questions you will ever ask.

You spend your entire life serving you. Buying things, fighting fights, arguing, loving, feeding, exercising, earning, and learning to satisfy needs for an image of you—a persona that doesn't even remotely resemble the real you. It's no wonder that your true needs are never satisfied, perhaps never addressed or even identified to start with.

The Illusion of Self is one of the deepest multilayered illusions humankind has ever needed to decipher. Philosophers, theologians, and psychiatrists have all tried to see through this illusion. And yet nearly all of us still wear masks on top of masks.

The illu-
sion starts with
a belief that
you are your
physical form.
A layer deeper,
you identify
with a persona
that is nothing
like you (your

ego), and then, at the deepest layer, you get deluded about your place in
the world. Like a Russian matryoshka doll, who you really are is hidden
beneath layers of illusions that should be uncovered, one by one.

When you uncover these, first you'll find out who you're not. Then
you'll keep shedding layers until you reach the one that's solid and real,
the one that will withstand the tests of *perception* and *permanence*.

The perception test is based on a simple subject–object relationship.
If you are the subject able to observe objects around you, then you are
not the objects you are observing. If you are looking at this book, then
by definition, you are not this book. The only way to see planet Earth
is from a vantage point outside it. Easy?

The permanence test, on the other hand, relies on a simple question
of continuity. If a quality or a description that you can associate with
yourself changes while you otherwise remain unchanged, then that
quality isn't you. If you were once a teacher and now you are a writer,
then those are changing states and neither is the permanent you.

In the previous chapter, in a stark departure from modern belief,
we established that your thoughts don't define you; you are not your
thoughts. This holds up to the tests. Your thoughts don't survive the test
of perception. If you are your thoughts—then how could you observe

them? They pop up in your head like images on a screen. You're not the image, and you're not the screen. The fact that you observe them is evidence that they are a different entity altogether. Neither do your thoughts survive the test of permanence: you don't stop existing in the few brief moments you manage to stop thinking. The moments when *they* cease to exist while *you* don't and the moments when *they* change while *you* stay constant are evidence that they are a separate entity from you. Thoughts fail both tests, and that's why they're not the real you. Let's apply those simple tests to the other identities people associate with.

This is a long chapter full of new ideas, so I suggest you get ready with a refreshing drink, a comfortable seat, and a clear mind.

Who Are You Not?

Your Physical Form

Before we get to who you are, it's easier to strip away the layers that you obviously are not.

Your body is the form that the whole world identifies as you. Your facial features, fingerprints, and DNA uniquely identify you. Everything you *are* is associated with this body. It must be you—it surely isn't anyone else!

But be honest now: have you ever looked in the mirror and felt that it wasn't you looking back? I have. I still do. Have you ever watched yourself on video and thought, *That looks weird* or *I really can't relate to the way I look*? Have you ever heard your voice on a recording? Did it sound like you? Until the day my publisher asked me to record the audio version of this book, I always thought I sounded like a little girl. They all laughed when I said that because, it turns out, I have a very deep voice. Even if you have not felt unrelated to yourself in that way yet, you surely will as you age or when your physical form changes while you continue to feel the same inside.

Think about the test of permanence. If the body you see in the mirror now is you, then who was it when you looked at your body as a six-year-old? Was that not you? What happens when you gain a few pounds? Is that *more* you? If, due to an unfortunate accident, one of your fingers were lost along with that unique fingerprint, would you no longer be you? Are the fingernails you snip off little pieces of you? What if you needed a kidney transplant? Do you become a bit of the donor and a bit of you?

Your physical body is made up of fifty to seventy trillion cells, and two to three million of them are replaced every second.[1] Red blood cells live for about four months, while white blood cells live on average for a year. Skin cells live about two or three weeks; colon cells have it rough—they die off after about four days. Your physical form is almost entirely replaced, sometimes many times over, every few years.[2] So which one of those ever morphing forms is you?

Think about the test of perception. If your body is you, then how can you see it and observe it? If it's the object—who's the subject?

This illusion takes only a few lines of text to see through.

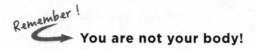

Remember! **You are not your body!**

Please take a few minutes to think about this and let it sink in. As you do, don't start thinking about who you are just yet. We're still discussing who you're not.

That body, though not you, takes so much of your attention. Many of us spend a lifetime caring for it. Tanning it, toning it, and tuning into it. Some spend a lifetime depressed because they want it to look different: taller, thinner, or stronger. Many take a tiny part of it—a nose, skin color, or a birthmark—and make it their tormentor every day of their life. Some cut it, stretch it, and stuff silicone into it. Some flood it with food and drink, while others deprive it of its basic needs in the name of a cult, a religion, or a fad. It always gets more attention than it deserves.

If you rented a car to go on a trip, would you start to believe that the car is you? If it remained with you for years, would that change anything? Your body is the physical avatar that takes you through the physical world, a vehicle, a container. Nothing more. That vehicle, however, is not nothing. It's important. If you were allowed to own only one vehicle your entire life, you obviously would take care of it, keep it healthy, in perfect working condition, and you'd make sure it didn't break or cause you troubles on your long journey. You would keep it looking clean and shiny and be grateful for the years of service and lifelong relationship it offered you. Still, whatever you did with it and regardless of how often you were seen in it, you would never think of it as you.

As if the illusion of the physical body isn't distracting enough, you distort things even further by adding more masks, until the real you fades beyond recognition. Let's follow the illusion but go a bit faster now. There's more. A lot more.

What Else Are You Not?

You are not your thoughts and not your body. What else fails the tests of perception and permanence? If we keep crossing items off, you'll eventually find the real you.

Maybe you are your emotions, as in *I am "in love."* That's a funny one. Who were you before you fell in love? What if your love grows? Will there be more you? What if it stops? Do you vanish? **You're not your emotions.**

Perhaps you are your beliefs: *I am a Hindu, a Christian, a Muslim, a Jew, an atheist.* Or *I'm spiritual but not religious.* What does that mean? If you adopt a new belief, does this make a new you? Who were you when you were two years old, before your belief systems developed? **You're not your beliefs.**

When asked who you are, you answer with a name: *I am Mo.* But obviously my name isn't who I am. Our names change into nicknames and married names, but we remain unchanged. **You're not your name.**

Some identify themselves with the group they belong to: *I'm Egyptian* or *I'm a fan of a certain football club.* But those temporary states change too. **You're not the tribe you belong to.**

I am the son of so and so. No, you're not. Those who discover that Mommy kept a secret and Daddy isn't really Daddy don't vanish from the face of the earth. *I am Tom's wife.* Okay, but who were you before you met Tom? **You're not your family tree.**

Then I must be my achievements. *I'm the inventor of this or the author of that.* Who were you before? *I'm a self-made millionaire.* What if all the money were lost, would the former self-made millionaire not be you? **You're not your achievements.**

I am the proud owner of a magnolia-black Rolls-Royce Phantom Drophead Coupe. Who are you fooling? Does the car you drive or the brand you wear determine who you are? If the Rolls-Royce gets stolen, does the thief become you? **You're not your possessions.**

You must be getting used to this approach now, so let's speed it up. You're not the bus you took to work this morning and not the bus driver. You're not the ant you just stepped on or the butterfly that took your breath away yesterday. You're not the pages of this book or the computer it was written on. You're not your cat or the sun or the atoms that make up our whole universe. Anything you ever observed isn't you, and everything that has ever changed in your constant presence isn't you either.

If you are none of the trillions of things around you, then who are you?

The Real You

It takes only one instant of full present awareness to meet the real you, one instant when you sit in silence and observe everything around you or observe what's inside you. Try to notice the thoughts in your head, the breath you draw, the feeling of your fingers touching the paper of this book. Try to feel the blood pumping to your feet. Sense the sounds around you, the light in your eyes. Try to notice the little details, the distant sounds of cars passing by and the smell of the dinner your neighbor is cooking.

You are none of what you just observed.

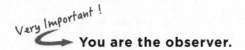

Very Important !

You are the observer.

You are the one aware of all that is happening around you. I know it may sound disappointing, but you have never seen *you*. You are not to be seen.

Very Important !

→ **You are the one who sees.**

I wish I could describe the real you to you in a way that your mind could comprehend, but unfortunately we're not fitted with the right equipment to do that. All of our human instruments are tuned to observe the physical world, and you—the real you—is not a physical object.

Tell

A quick game of Tell can help you understand why it's so difficult to describe you. Let's start with an easy round. Please tell me what the ocean looks like two hundred feet below the surface and two hundred miles off the southeast coast of New York. Can you tell for sure? No you can't—because you've never seen it.

Okay then, can you please tell me what's playing on the broadcast radio waves around you right now? Without a radio? Can you? Of course not. Even though the waves are all around you, they're not perceivable because you don't have the instruments to receive them.

Finally, here's a tricky one. Would you please tell me how the smell of freshly baked cookies looks? Can you? Why not? Because a smell is not a visual property—it can't be seen.

Now go ahead and tell me how the real you looks. Can you? Like the depth of the ocean, the real you is something you've never seen. Like radio waves, you don't have the instrument to perceive it. More importantly, because of its nonphysical nature, it is not to be seen. Being seen is a characteristic only of the physical world.

The fact that you can't comprehend your true nature doesn't mean it doesn't exist. The depth of the ocean, the ubiquity of radio waves,

and the smell of cookies all exist regardless of your capabilities to fully comprehend them.

To stretch our imagination a bit further, consider this: to observe the physical
world, you need to observe from a vantage point outside it. (Similar to how you
can observe a building only when you
step outside of it.) This is an idea with a
lot of creative potential. Movies such as *The Matrix* and *Avatar* flesh out
this concept to the maximum. The characters in those movies control
their physical form from afar. Likewise, in your life, if you're not part
of the set in which the scenes of your life unfold, or even identified
with the body of the avatar you use to navigate them, then the reality
portrayed in those movies may not be far from the truth.

Let's pause for a moment to think about what we just discussed. I
found it to be a truly cool concept when I first understood it. Here's
why. When you're not your wealth, then a lack or an abundance
of money can't touch the real you. If a thief took off with some of
your hard-won earnings, it might affect your physical form and your
thoughts might make you suffer, but the real you would lose nothing.
You would be observing those changes while you remain unchanged.

The illusion driving you to protect all of the possessions in your
life is an attempt by your physical form to control the physical world
around it. The real you is unaffected by that physical layer and all that
it contains. A sudden loss of status, for example, wouldn't bother you
as much; you would be identifying with the real you and not the temporary illusion of you. With no fear of loss, then, and with no worries about the future, you would understand that nothing could really
hurt you.

Now let me provoke your thought a bit further. We've established that you are not your aging, changing, mortal body. Imagine, then, that you are in an accident and lose both your arms and legs. Your physical form will be reduced by half. It will be suffering and may change your thoughts and behaviors. Yet you will not lose half of the real you. Your real self will remain unchanged.

Now take this concept to its extreme. Imagine that you lost 100 percent of your physical form and ask the interesting question: What happens to the real you when your whole body is lost? Does it stop existing? When your body dies and decays, where does the real "you" go?

In my personal belief, the answer is *Nothing will happen to "you."* You will just stop connecting with your physical form. You will still be you, and you will say, "Wow. That was fun!"

This belief helps me realize that Ali's physical form may have decayed, but his true amazing self still lives on—and that one day I too will leave my physical form behind and that will be okay. It's a wonderful thought in such a difficult time.

I know this is a leap of faith for those who strongly associate with their physical selves, but once you truly see, you'll never be able to go back. To reach the state of uninterrupted joy, you need to accept that everything in the physical world will eventually vanish and decay, but the real self will remain calm and unaffected. Connecting to that real self to see through the illusions of the physical world delivers the ultimate experience of peace and happiness. Let's keep exploring. It will all come together nicely in the end.

Who Do You Think You Are?

In trying to establish who we are not, we uncovered a lot of masks that we wear to create an identity. Those masks represent the next layer of the Illusion of Self. They can all be summed up in one word that has tormented mankind since the day we became a society.

Your Ego

That word is *ego*.

Ego is not used here to mean arrogance but rather to mean a sense of identity, a persona—the way you see yourself and believe (or wish) that others see you.

We are all born without an ego. We start life with no sense of our own selves as separate entities from the rest of the world. We spend our few waking hours fully immersed in the present moment. As we start to play, we calmly pick one toy, then let it go to pick up another without a single negative thought in our head. The serenity is temporarily interrupted when we're hungry or Mommy goes out of the room, but once the annoyance is removed, the calm returns.

The next stage of development, however, brings a fundamental change. It all starts when you notice that Mommy, or whoever cared for you at the time, associated names with things. She referred to herself as "Mommy," to your toy as "Dolly," and to you using some kind of cute name, say, "Pooki."

As soon as you manage to control your speech processor to produce your first word, you blabber out a name: "Ma-ma." Mommy runs to you laughing, hugging, and kissing. "Yes, baby, I am Mamma. I love you and will run to you every time you call me."

Well, that's interesting, you think. The overwhelming excitement you receive as a result of that first word teaches your brain that naming things gets you praise, so you accelerate the process. *Tlee, Tat, A-plen, caar,* and *mik.* Because the way you say those things is so cute, it gets you more praise, and so you expand your vocabulary further until you say the word that will forever change your life and become your identity— and the central focus of your brain as long as you live: "Pooki!"

Judging by the way things unfold from then onward, that moment must rank up there as one of the single most important moments of your life. You have an identity. Initially you refer to yourself in the third person: "Pooki hungry." Then Pooki becomes "I," turning that single letter into the absolute center of your entire existence; soon after, you add "me," "mine," and "my," and the process is complete. Your ego is born.

Pooki becomes possessive. You associate yourself with objects in order to create a more comprehensive identity. The innocent child who had been completely happy to play with anything starts to have a favorite toy. "*My* Dolly." And if "my Dolly" disappears, Pooki will feel pain and cry. Playtime becomes as much about building your identity as it is about playing. Certain toys are now needed to make you happy not because they're more fun to play with but because they're part of an identity that makes you feel complete.

It gets worse when you learn to compare your identity, made up of I, Me, My, and Mine, to the identities of those around you. Being "less" than others hurts you. Even if you have your favorite toy, not having a toy that your friend has makes you feel inferior to your friend. You start to dislike your toy; you ask Daddy for the other toy; and you complain if he says no. You beg and plead and finally get it, only to give up on that new toy as quickly as you wanted it when the next thing comes out.

What happened to the joyful, calm infant who simply enjoyed the moment with whatever it had to offer? Gone. Swamped by the constant urge to define an ever-evolving identity.

Things get even more interesting when the brain moves beyond the physical world of toys into the intangible. The simple act of letting go of the table to stand on your own, and then taking one step forward before falling on your bum gets Mommy all excited. She shouts, "Bravo, Pooki!," runs to pick you up, kisses you. She's happy and laughing as if you conquered the world. And you think, *Well, that's something. Maybe I should do more of those simple tricks to get more of that amazing appreciation and attention.*

For weeks you call out, "Mommy, look: Pooki walk!"

"Yay!" she shouts.

"Mommy, look: Pooki stairs."

"Wooohooo!" she screams.

"Mommy, look: Pooki found a toy!"

"Bravo, Pooki!"

"Mommy, look: Pooki pick nose!"

"No, no, Pooki. Bad Pooki."

Hmmmm. You realize that certain acts are socially acceptable; they get you praise and encouragement. Other acts are frowned upon. Smart as you are, you do more of the former and less of the latter. You start to build a persona, an image of how you want to be seen in order to fit in and be accepted. It no longer matters what or who you really are inside; what matters is what you appear to be. Your attention, for the rest of your life, shifts away from your reality to your image.

Your addiction to maintaining your image is then intermixed with an addiction to attention. You quickly realize that picking your nose gets you more attention, while putting your toys away neatly goes unnoticed. And attention is what you seek. The rebel is born. The "attention fighter" takes over. I shall be noticed, it says, whatever the cost.

Those identity crises intensify in our teenage years, when our insecurities and pressure to fit in are at their peak. We get further and further away from our *true* nature and closer to the *accepted* nature of our peer group. If having sex at fourteen gets us accepted, some of us will jump in. Bravo, Pooki. If playing football is cooler than joining the science club, forget the science club. Bravo, Pooki. And if staying away from drugs, cigarettes, and alcohol makes us seem lame, bring 'em on. Bravo, Pooki.

And then adulthood. We go to work, dress fancy, and repeat meaningless words: *synergy, chasm, ecosystem, spin-out,* and *drink the Kool-Aid.* What kind of language is that? It doesn't even sound like English, but we use it because it gets us accepted. We become serious and make sure we don't show our emotions at work. Some of us learn golf, go to business dinners, and attend the team party. We fit in. Bravo, Pooki. As we get ahead, some of us invest in brand-name clothing or expensive cars to keep growing our persona. Everything we earn from the persona is spent on maintaining it, but none of it makes us really happy. Yet, we never stop to reconsider what we do as long as it keeps our egos intact.

The Roles We Play

Once we start wearing masks to reinforce our egos, we spend the rest of our lives playing roles. There is the role of a powerful executive: professional, well-dressed, composed, heartless. The role of a mom: using baby talk with her kids, dressed in sneakers, and going to coffee mornings with other moms. The role of an artist: eccentric style, rebellious and mysterious. The role of a patriot: proud, signed up, and supportive of killing the enemy. The role of a sophisticated gentleman: always critical, seemingly knowledgeable about art and culture, pausing for a moment of silence and deep thought between every two sentences. The role of a seductress: sexy dress, high heels, and a dreamy voice. The role of the macho bad boy: tattoos, self-confidence, and a careless demeaning stare.

Every identity becomes a role, even the most basic ones. Boy or girl, for example, starts a lifelong stream of behaviors based on societal expectations. Pink or blue, dolls instead of a football, skirts instead of shorts. The role becomes your expected image, and if you feel different from what the herd expects from you, life can be a struggle.

There are the roles of old and young, led by the societal expectations that when we get older we should behave differently than when we were youthful and playful. When it's time for children to face the "real world," everyone demands that they become "serious." They are asked to sit still for hours at school—no talking, moving, or playing. We prioritize homework over discovery and adventure. We expect children to stop being frivolous and start to be punctual and conformist. Some resist for a while, but sooner or later most comply.

What would happen if we all took off the masks and did the best work we could without pretending to be something we are not? Would we close fewer deals or invent less? I don't think so. The work we do, not the mask we wear, is what pushes us forward. In an ego-less world where it doesn't matter how we're perceived, we would dedicate ourselves to doing our very best and aiming for the best results regardless of how others perceive us. While the ego of a professional is built on making the job look hard, often the best results are achieved by doing very little. The best managers, for example, hire talented people and manage sparingly. As the need to pretend is removed, the best professionals often turn out to be those who don't play the role at all.

The Masquerade

For each role there is a look, a dress code, a lingo, a peer group, an enemy to hate, trendy topics to discuss, facial expressions to fake, and common sorrows to worry about. It's easy to learn the image. It's on TV every day. Cut and paste, and we all become actors. We wear different masks and hide our reality from everyone, including ourselves.

Our assumed identities becomes our whole lives, and we start to believe them—even more than others do. They often sense anomalies in our behaviors. They compare the roles we're playing to their corresponding images from the media. They sense that they're an act and eventually reject them.

When our self-image is attacked or threatened in any way, our instinct engages to protect our ego. Our instinctive fight response makes us argue and quarrel, and our flight response makes us withdraw and become depressed. Those primitive caveman tools have evolved to fit the modern world of ego. The spear became brand-name clothing and expensive cars. The hunters' gestures became slang, and our best camouflage for fitting into our environment became Facebook likes. Through it all the Happiness Equation malfunctions completely because our expectation that others will buy into our fake image is never satisfied—and we feel unhappy.

I can relate to this firsthand. I lived the experience at the peak of my depression. For years I was obsessed with cars. Their artistic engineering intrigued me, but more importantly, they served my ego. I chose the persona of a successful sophisticated collector and was miserable living it. While I still love cars today, I've lost the urge to own them. I realized that my passion was contaminated by the urge to satisfy my ego. Before I became successful, the cars I bought were a lie to pretend and cover up the fact that I *hadn't* yet made it. And when I truly became successful, I didn't need a car to prove it. In both cases, cars didn't make me happy. No prop for an ego ever will.

Arabic folk culture tells the story of an old teacher who's visited by many of his students years after they left his class. They talk about how successful they've become in life and shower gratitude on their beloved teacher. Then they all start to talk of the pressures they're facing, the stress they go through to keep up with expectations. Success is not making them happier.

The teacher gets up to prepare a large pot of coffee and comes back with a tray that contains a variety of cups. Some are made of crystal, some of silver, and some are cheap plastic cups. He asks the students to pour their own coffee. They all reach for the most beautiful and expensive cup available.

When they sit back down, the teacher acknowledges the most beautiful cups, but then points out that what they all really wanted was coffee. Regardless of the cup, the coffee was the same. If social status, fashion, image, possessions, and social acceptance are like the cup, he says, then life is the coffee. Why do we try so hard to drink from a fancy cup when all we want is good coffee? If you want to live a stress-free life, he says, ignore the cup and just:

Remember! **Enjoy the coffee.**

The Darker Side of Ego

Ego isn't always about vanity. Often the images people build for themselves are negative. They believe deep inside that they're worse than they really are. The "victim," for example, is a very common ego type. You think the world is always against you and that you're bound to suffer. When your ego is threatened, you become offended. "What do you mean it's okay?" you'll say. "Nothing is ever okay. I've earned the right to be unhappy and paid for it with my painful experiences. If suffering is a choice, then I chose to suffer. It is who I am."

Negative personas can be driven by feelings of diminished self-worth, self-pity, guilt, or shame. *I'm fat. I'm ugly. I'm short. I'm stupid. I don't deserve to be loved. I'm a sinner who deserves to be punished.*

Another of the common negative roles people play is the role of a grieving parent. One of the first thoughts that rushed to my mind when Ali left was *Death took away my son.* I could have acted like the victim. It would have been easy to fall into this trap because Ali wasn't only an integral part of my life but also a pillar that supported it. I can't even remember life clearly before I assumed the role of Ali's father. This role is a tricky illusion to see through, one that leaves many of us stranded in suffering for countless years. The thought, however, is false. Ali was never *mine.* Ali was his. He led a life that took him places. Sometimes I was a part of his story and sometimes I wasn't. When he needed to leave home and travel thousands of miles to attend Northeastern University in Boston, as much as it pained me to be far away from him, I supported his choice. I was happy that he was following his path because it was his life and not mine. Now that he jumped onto a whole new path, why should I react any differently to his death? While I will always miss him, I know it's his path. He was never mine.

Many of us go down a sad path when we let our egos make us suffer. What happened to little Pooki, the only true identity you ever had? Calm, happy, totally in the moment, sitting naked in a diaper without a care in the world, no sense of self, uncluttered by the thoughts of "I." No thoughts of how I look, what I represent, what people think of me, or even what I think of myself. Pooki was happy with whatever came, not possessive, ready to let go, and ready to move to the next toy without attachment.

Don't you wish you could get that back?

Well, it never really left. *You* never left. The egoless child is still calmly sitting inside each of us. Buried in layers over layers of lies, egos, and personas. Happy nonetheless. Waiting to be found.

Let's find your Pooki.

Undress

Like a Russian matryoshka doll, you'll need to remove the layers one by one, trying to distinguish the real you from the roles you assumed over the years, until you find your pure self. Until then, *undress*. Remove all the masks of the ego.

When I say "undress," I mean that quite literally. This exercise might be a bit shocking, but it's very effective. When you go home tonight, close your door and in the privacy of your room stand in front of the mirror. Look at everything you're holding, using, or wearing. If any of it extends beyond its basic utility, take it off. It's just there to serve your ego.

Look at that shirt or jacket or dress. Did you buy it just to cover you and keep you warm, or does it also serve to help you create your self-image? If you didn't want to *look* pretty, elegant, carefree, or artistic in front of yourself and others, would you have bought something different? Check out those jeans. If you didn't want them to make you *look* sexy, would you have bought them a size bigger? How about your shoes? If you didn't want to *look* professional, would you have bought something more comfortable?

Look at your jewelry. Does it serve any real utility at all? Does it serve you in any way other than the image it portrays? Are you wearing a ring because a loved one gave it to you or because you want to tell the world that you are loved? Would you have bought a different watch if all you wanted was to tell the time? If any of those accessories are there for their pure utility, keep them on. Otherwise, undress. And put them all away.

Look at your makeup, your nail color, your haircut. Do any of those serve any real utility? Look at that tattoo. Was that really because you wanted to cherish a memory, or did you want to be *seen* to be cherishing that memory? Even if you can't remove the tattoo physically, remove it mentally. Take off the urge to send that message or build that image for the rest of the world to see.

See how much we put on every day to serve nothing but our ego? See how little is left to carry around if you strip off all the images that you constantly work to maintain? See how light you feel without them?

Now look at that bare body, stripped of all the ego accessories. You're back to that little Pooki with nothing on but a diaper. Now we can go even further. Whether your body is toned or overweight, ask yourself, "Does that make me fit in one role or another?" Have you been exercising to stay healthy or to look athletic and attractive? If it was just to stay healthy, would you have exercised differently? Is that body even you? The muscles, hair, blood, mucus, and sweat—is that you?

No, you are the one observing it. The one who would remain aware even if you gained or lost a hundred pounds. That pure one inside is Pooki. You found it. **Bravo, Pooki!**

A Losing Battle

Trying constantly to get approval for your chosen image is a losing battle because the real you isn't what the ego pretends to be. This makes us unhappy since we're always searching for the next thing to make the image complete in the hope that people will believe that's who we are. This will never work, for two reasons.

First, others will rarely ever approve of your ego because they are more concerned with their own ego than with yours. The survival of their ego depends on comparing it with yours. For them to be right, you should be seen as wrong; when you're less, they

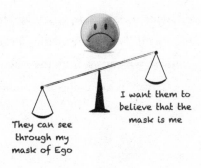

I want them to believe that the mask is me

They can see through my mask of Ego

become more. Disapproving of someone else is the easiest way to feel superior. It doesn't require the hard work needed to become better. It just requires thinking less of someone else.

Everyone does it. Some are silent in their judgment; some are loud and public, criticizing or cursing. People will disapprove of you not because they are evaluating you but because they are valuing themselves. There's no way to win. It's sad but true.

A fable captures this in the story of a man and his young son traveling to the market. They had only one donkey, so, out of respect, the son suggests his father ride. A passerby comments, "What a cruel father. How can he ride and let his young boy suffer?" To gain the acceptance of others, they switch: now the boy rides and the father walks. In no time they overhear another passerby: "What a disrespectful son. How can he ride and let his elderly father walk?" The son then suggests to his father that they both ride, only to hear "What a cruel family. Have they no mercy for this poor donkey?" So they decide to show that mercy by carrying the donkey to the market, where people call them crazy and kick them out!

The second reason trying to get others' approval will always fail is because they won't be approving of *you*. They'll be approving of your persona. It'll no longer be "Bravo, Pooki!" but more like "Bravo, unrecognizable ego that looks vaguely like Pooki!" And you'll feel it. You'll feel, deep down inside, that your effort was spent to gain praise for someone else. It'll make the victory feel empty and make you feel that the real you isn't worthy. What's the point in spending so much effort to gain acceptance for someone other than you? Wouldn't you rather serve you?

I am me

I'm surrounded by those who love people like me

Ali had an amazing talent for *not* seeking acceptance. It always made him calm and confident. He was the happiest young man I ever met. He was never apologetic about following his own rhythm. In a very smart way, he believed that instead of pretending to be something he wasn't in order to please everyone, it would be easier to find those who liked the type of person he really was. First, he was very selective about who he let into his life, and when he found the right fit, his friends loved the real him. That gave him the confidence he needed to be himself. Later, he opened up and let everyone in, but he made it clear that "what you see is what you get." They were all drawn in by the light of his genuine pure self shining through.

You'll never please everyone. Find those who like the real you and invite them closer. All others don't matter to you.

Ali's wise mother often recited to him a line from Sting's song "Englishman in New York":

Very Important!

Be yourself no matter what they say.

More important, *Love who you are.* The real you is wonderful and calm, just like Pooki. The versions of you that you don't like are actu-

ally those personas created by your ego. You are all you'll ever need—and all you will ever have.

Undress. Drop all the other copies and love the real you. Bravo, Pooki!

Now would be a very good time to pause and reflect. We still have one more layer of the Illusion of Self to uncover. It requires a clear mind, so please take your time.

The Star of the Movie

Your Place in the World

Perhaps the deepest part of the Illusion of Self is the part that causes us the deepest sorrow. It is the part that most often prevents us from solving the Happiness Equation correctly. It begins when you believe that you are the center of the universe, that good things happen because you've earned them and bad things happen just to annoy you. And that's the furthest thing from the truth.

Let's dive into the thoughts of Tom. We'll dive inside his head on a

Saturday morning, while he quietly enjoys a cappuccino in front of the wonderful view of the San Francisco Bay Bridge.

This must be the best coffee ever, he thinks. *The barista paid so much attention to making it—and then she added this wonderful art on top. She must know how much I love a good coffee.*

The experience reminds him of Tammie.

She showed me this place the first weekend I decided to move over here to be with her. That's the same day we bumped into Timmy. We chatted about the good old days before he mentioned that the start-up where he worked was hiring. He sure did me a favor by recommending me for this job. I like Timmy. Tamo, my first boss, was so tough, but he taught me so much. I don't really know if I should like him or hate him. He was so generous with the stock options too. That made a big difference to my start here. Of course, I'll never forgive him for cheating on me with Tammie, but hey, that worked out well too. I'm much happier with Tamar.

I'm so fortunate to have so many supporting actors in my life. If this life was turned into a movie it would be a love story of suspense, action, and drama. Everyone plays a small role that all leads to me, sitting here now to enjoy this cappuccino. I like my movie. It comes with its difficult moments, but I get it. How else could I be the star if I didn't have to battle a few challenges and come out triumphant at the other end? It must be an important movie for so many actors to support it. I must be destined for greatness. This is the movie. Well, it sure feels that way to me.

Remember!

→ Well, it sure does feel that way to all of us!

Let me ask you a question: If you are the star of your movie, then who is the star in Tammie's? If she is, then what does that make you? A supporting actor, perhaps?

Take this logic a step further. You're featured as a supporting actor in Timmy's, Tom's, and Tamar's movies. You just supported the movie of the barista, the lady you held the door for, and the poor ant you squashed as you walked into the building.

If you're the supporting actor in thousands of movies but the star of only one, how much of a superstar does that make you?

Have you ever considered that your behavior may have been what led Tammie to cheat? Maybe your behavior also led to her life of tension and unhappiness with Tamo, which led him to cheat on her, which massively affected the life of their daughter as they separated?

And did you even notice that you changed your mind in the last split second when you were stepping into the café? You decided to go for the barista on the left. That made the person behind you go to the other barista in a life-changing coincidence, in which they meet for the first time, fall madly in love, get married, and have a child who will grow up to become a famous surgeon and will save the life of your granddaughter forty years from now?

Did it even cross your mind that the $2 tip you gave the taxi driver helped his father buy the organic fertilizer he needed for his coffee plantation, the same one that harvests that incredible free-trade coffee you'll enjoy the next time you come to the same café? Perhaps the profit the farmer made will save the life of his child, who will become a mad scientist and end civilization as we know it. You'll never know. Your movie is just one of many entangled movies. Billions, in fact!

You live in a complex web of connections. Every single day, every step you take and every move you make impacts—even if in small ways—the life of everyone around you and perhaps occasionally the life of everything on this planet. This happens while every step any of them takes might affect you.

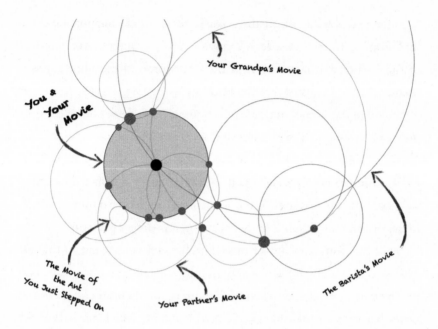

Your Grandpa's Movie

You & Your Movie

The Movie of the Ant You Just Stepped On

Your Partner's Movie

The Barista's Movie

Good versus Bad

This very intricate web of lives forces you to confront an unfamiliar concept: good is never all good, and bad is never all bad. The vantage point of the story may render a very different impression. And as we've just found out, there are countless vantage points.

A bad event for the main character of one movie can turn out to be the *best* event for another character in an overlapping movie. Like the old saying: one man's loss is another man's fortune. Allow me to use an extreme personal example.

Ali was fasting, observing the month of Ramadan, on the day he left us. Upon his admission, the hospital asked that he keep fasting in preparation for his operation, which meant that his last sip of water before he left our world was almost a full day before his departure. His day of thirst moved me and many of his caring and generous friends, devastated by the shock of losing him, to donate in his honor to the noble cause of providing fresh drinking water to poor communities around the world. Thousands who suffer from the lack of this precious necessity were nurtured as a result. Perhaps as I write these lines you turn your attention to the cause you might donate to as well, and so together we would reach millions. Now the question becomes: Was Ali's suffering good or bad?

Well, it depends on the vantage point. To him it was bad. The physical pain of thirst for a full day must have been harsh. And yet Ali lived to help others. Given the chance, he would have surely volunteered to stay thirsty for a day if it meant that thousands of others could drink for life.

To me, a loving father, his suffering was horrific. But I take comfort in knowing that he would have liked the next scene of the movie. It would have brought a tear of happiness to his eyes that others were helped.

For the thousands whose lives have changed, and who don't know the cause of their blessing, this was surely a happy time in their movies. The truth is:

Very Important !

Everything is both good and bad.
Or perhaps everything is neither.

Even at the individual level, with the passage of time nothing is all bad. How often did something in your own life start out as bad but turn out to be good? That muscle pain after your latest jog might help prevent your heart attack twenty-five years in the future. On the other hand, perhaps the pleasure of driving your car very fast might, a few seconds later, make it the last car you'll ever drive.

Widen your vantage point to see the same event from different angles. Buying that new car is good, but parting with your money is bad. The pain of touching a hot iron is bad, but saving your fingers is *very* good.

Good and *bad* are just labels we apply when our minds fail to grasp the comprehensive, never-ending movie spanning across the billions of lives and extending over all of time. If we could grasp the complexity of the web of perspectives that compose our experiences, we would realize that everything is just what it is, just another event in the endless flow of the big movie that features all of us.

We will cover the concepts of good and bad in extensive detail later. For now, look beyond a single vantage point of a single snapshot of your own individual movie, and you will always find the good within the bad. Every event will contain something that meets your expectation and make your Happiness Equation work. That optimistic perspective will make you happy. Our ego makes us go through life feeling that life is all about "me." Life gives *me* or takes away from *me*. You might think that the traffic on your daily commute or the line at the supermarket checkout is there just to frustrate you. You might think the universe took the effort to get the road built and cars invented, get so many of them sold and have all of the drivers summoned up in that one road on that specific morning to annoy you. Of course, how else could it be when you're the star of the movie?

You are one of more than 7,200,000,000 people. One of trillions of other beings living on a massive planet that is tiny in comparison to the size of a solar system, which is a negligible part of a small galaxy, of which there are billions spread across our infinite universe. Every living being, atom, and beam of light is diligently following a path that occasionally happens to overlap with yours.

Get real. You are not the star of the movie. Most of what happens around you isn't about you at all. There are infinite numbers of other movies. In those, if you feature at all, you're just a supporting actor. It would really help your happiness if you started to look at your life that way. Look at the night sky and remember that its beauty resides in its billions of sparkling stars. Of those billions, you are but one.

Remember!
You are not the star of the movie!

Chapter Five

What You Know

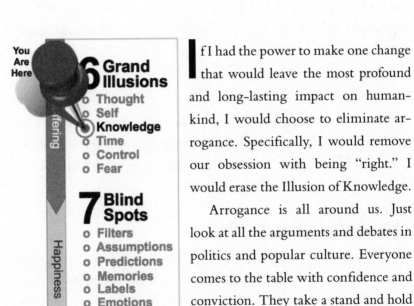

If I had the power to make one change that would leave the most profound and long-lasting impact on humankind, I would choose to eliminate arrogance. Specifically, I would remove our obsession with being "right." I would erase the Illusion of Knowledge.

Arrogance is all around us. Just look at all the arguments and debates in politics and popular culture. Everyone comes to the table with confidence and conviction. They take a stand and hold their positions firmly, confirming what they know. Their confidence seems convincing—but do they really know?

Our pursuit of knowledge has propelled our civilization forward. It's moved us from chipping stones by the light of a bonfire to rushing around city streets talking into our smartphones. Knowledge is the fuel of civilization. But at the same time, our conviction that we truly *know* causes us to suffer. It's the ultimate ignorance. Before we discuss how this affects our happiness, let's first assess the magnitude of the illusion.

The Interview

If you were asked to interview someone who claims to be knowledgeable, you would ask questions with the aim of uncovering the depth and breadth of that knowledge. You would try to assess how accurate her answers are and how much she knows about the subject in comparison to all there is to know.

If she knows a lot and her information is accurate, she will be considered an expert. If, however, she knows very little and much of what she knows is wrong, you will dismiss her claim of knowledge—and politely ask her to leave. Well, let's go ahead and interview humankind (including me and you). Let's see how much of an expert it really is.

The Depth of Knowledge

What matters most isn't what you know, but how accurate your knowledge is. To know the wrong thing is worse than not to know at all. Correct?

At a news briefing in February 2002, U.S. Secretary of Defense Donald Rumsfeld was asked a question about the intelligence surrounding hypothetical Iraqi weapons of mass destruction, the supposed existence of which was the reason given for starting the war. Cryptically he responded, "Reports that say that something hasn't happened are always interesting to me, because as we know, there are known knowns; there are things we know we know. We also know there are known unknowns; that is to say we know there are some things we do not know. But there are also unknown unknowns—the ones we don't know we don't know. And if one looks throughout the history of our country and other free countries, it is the latter category that tend to be the difficult ones."[1]

The consequences of that last category have exacted quite a toll—heartbreaking, really.

Shockingly, the accuracy of most knowledge—even scientific knowledge—suffers because we ignore unknown unknowns. Take physics, for example. Sir Isaac Newton discovered gravity and published his laws of motion in 1687, forming the foundation of what we now know as classical mechanics. Those laws were fiercely debated until they were undisputedly proven and accepted. Once proven, scientists embraced them as facts that govern everything from the falling of an apple to the orbiting of the moon and planets. Anyone who dared dispute their accuracy was considered ignorant. The arrogance of debate was replaced with the arrogance of absolute knowledge. This position, however, was totally unfounded because Newton's laws ignored many *unknowns* that were later discovered.

In 1861 James Clerk Maxwell's classical thermodynamics rendered Newton's laws insufficient. In 1905 Albert Einstein declared Newton's assumption about time to be false. In the mid-1920s, quantum physics showed that the world of small particles doesn't behave as Newton expected. In the 1960s string theory exposed the incompleteness of quantum theories, which, in turn, was rendered incomplete in the 1990s by M-theory—and it seems just about time for some other new term to render *that* incomplete very soon.

Can you see how misled we can be? Something as basic as the elementary laws of physics that seemed to function properly and accurately for more than two hundred years was, at best, an approximation.

DDAA

In the modern world, our access to knowledge has exploded. Every answer we seek is just a search term away. Billions of pages populate the web, ready to answer any question you may have. It's hard to imagine that there's anything we humans don't know. But don't be dazzled by the large numbers. The real question is, How much of it is accurate, and how much is just a claim of knowledge? The reason you get millions of results for every search is because every topic is presented from countless points of view. Some are vetted by the wisdom of the crowd to be more relevant, but no one can confirm beyond doubt that what you read is true. Every question you'll ever ask will be governed by a refinement cycle that I call DDAA: Discovery, Debate, Acceptance, Arrogance.

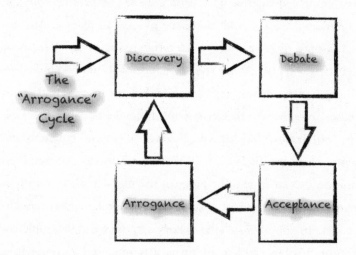

For thousands of years, humans have pondered questions about our universe: questions about who we are, what we're doing here, and how everything works. Every now and again we stumble upon incredible *discoveries*. The new knowledge drives *debate* and disagreement until one side is proven right by what seems like undeniable evidence. This leads to *acceptance* of the new knowledge as fact. Comfort with our knowledge inevitably leads to periods of *arrogance*. We think that our knowledge is confirmed beyond any doubt and argue fiercely with those who contradict it, only to realize—in the next wave of discovery—that what we know is not complete and sometimes not even correct. This cycle—DDAA—has always been our journey, with ever-incomplete and inaccurate knowledge.

The reason we so arrogantly believe in our knowledge is that our observation often validates it. Our ability to navigate our immediate physical surroundings, for example, was never affected by our false assumption that the Earth was flat. It's hard to imagine something new until new observations contradict our prior understanding—seeing the hull of a ship vanish before its mast in the horizon, for example. Only then do we revisit what we know and even start to wonder how we ever thought as we once did. How could we miss what now seems so indisputably obvious?

The kind of knowledge that turns out to be incomplete is the illusion that we live with every day in science, politics, history, and even in our personal lives. You might think someone is snobbish, only to find out that he's actually just shy; you expect the bank to help you out, only to discover it's ripping you off; you anticipate a pair of shoes making you happy, only to discover they hurt your feet. Even our dietary habits suffer: the accepted wisdom about which vitamins and minerals are good for us keeps changing as scientists flip flop and tell us to stay away from the things they were encouraging us to take just a few years earlier. It's all an endless loop of DDAA! Discovery leads to debate, then to acceptance, and on to arrogance—which is then diminished by new discoveries.

Through it all, we humans continue to assume that we possess the ultimate knowledge. We behave as if we, the smartest beings on the planet, know it all. We reject any possibility that something could be missing, let alone wrong.

The Breadth of Knowledge

Even in the few instances in which we know something accurately, all that we know is truly insignificant compared to all that is out there to know.

For example, the universe is made up of more than 96 percent dark matter and dark energy, the transparent stuff we previously called a vacuum and about which we know very little. Down here on Earth, more than 90 percent of the oceans' volume remains unexplored. A Godzilla could be swimming there as you read this, and we wouldn't have a clue. Even inside our own bodies, we understand the purpose of about 3 percent of our own DNA—so we call the rest "junk DNA." We call it junk because we're too arrogant to admit that we just don't understand what it's for. Every day new discoveries are made that help us understand more. But until we get the details of it all, the humble thing to do would be to consider humankind at least 90 percent ignorant. So much for knowledge!

The breadth challenge isn't limited to science. It extends to every part of our lives. How much do you know about what's going on in your friend's life before you feel upset that he didn't return your call? How much do you know about the hardship a shop attendant is enduring before you judge her for not smiling back at you? How often do you decide to follow a diet being presented as the new revolutionary discovery when you really know almost nothing at all about how your body really works?

Because we truly know very little. Yet to summon the conviction we need to believe in our actions, we convince ourselves that our knowledge is complete when, in fact, so much is missing.

What's Missing?

It's not just arrogance. Sometimes our knowledge is restricted at the most fundamental level, at the level of our senses and by the basic building blocks we use to form thoughts and concepts.

Our Own Senses Are Limiting

Even when it comes to our own senses, we're arrogantly certain of what we observe, though it's clear how little we can rely on our per-

ceptions. When we touch a rock we feel it to be solid, while in reality it's almost entirely made up of empty space. We miss out on hyper pitches of sound that dogs can hear and all the infrared light that mosquitoes, fish, and some snakes can see. If you live in Moscow your perception of "cold" will be very different than if you were born and raised in Dubai.

Are you sure what you see here is a horse—or is it a frog? A simple tilt of your head renders a different reality, one that seems so real it's hard to deny or force your brain to refuse it. There must be trillions of bacteria around you, affecting you as you read. There are more bacteria in your own body than there is "you" in it, but you don't see them. If you've ever spoken to someone who's color blind, you realize that he functions just as efficiently as you do even though his vision of the world is drastically different from yours. It's safe to assume that even when it comes to our very own perceptions, we don't "know" for sure.

Our Words Are Crippling

Another fundamental limitation on our knowledge is to be found in the very building blocks we use to think and communicate. We use words to define concepts, but there's no way words can encompass them fully. Take, for example, *mango*. The word is a mental construct that refers to "a juicy, fleshy, marigold-colored, aromatic sweet fruit." The word helps you understand what I'm referring to, but does saying *mango* create the experience of smelling or eating one? Do the words *fleshy* and *aromatic* give any accurate knowledge of what it's like to sink your teeth into a juicy, ripe mango and enjoy the intense sweetness, the flavor, and the aroma? Can you use these words to construct the difference between a mango and a peach, especially if you've never experienced either?

The limitation inherent in words extends to all aspects of knowledge. There's a delightful color we all agree to refer to as "sky blue." And yet there is no way to prove that the visual image you get when you see the color of the sky is the same as the one I get. Language can't help us synchronize that understanding. For all we know, the way you actually see sky blue could visually appear to you as the way I see rose pink. We both agree that it's a delightful color, and we agree on its wavelength and name, but we can never know if we actually *see* the same thing.

We distort knowledge further when we abstract layers of complexity into one simple word. *Skyscraper* packages countless sophisticated designs, thousands of different materials, and the work of millions of people into one extremely terse description. It mistakenly suggests that all skyscrapers, at least at some level, are the same. Once we have a word to describe a concept, we assume that we know that concept regardless of how superficial our knowledge really is. How can we pack concepts such as love, devotion, divine, or society into a single word? Think of the magnitude of knowledge that we attempt to pack into the words *philosophy, psychology, sociology*. Are all atheists or pragmatists so similar that a single word can describe all of them? How accurate a description does the word *death* offer compared to the concept it describes? Do the words *autocracy, meritocracy*, and *democracy* capture what they refer to? Are they applied as described?

In his rude yet very funny movie *The Dictator*, Sacha Baron Cohen plays a Middle Eastern dictator who is being compelled by events to turn his country into a democracy. He gives a speech describing the benefits of a dictatorship:

> Why are you guys so anti-dictators? Imagine if America was a dictatorship. You could let one percent of the people have all the nation's wealth. You could help your rich friends get richer by cutting their taxes and bailing them out when they gamble and lose. You could ignore the needs of the poor for health care and education. Your media would appear free, but would secretly be controlled by one person and his family. You could wiretap phones. You could torture foreign prisoners. You could have rigged elections. You could lie about why you go to war. You could fill your prisons with one particular racial group and no one would complain. You could use the media to scare the people into supporting policies that are against their interests.

Is he confused about the meaning of the word *dictatorship*, or are we confused about democracy? Or are all words, perhaps, being used loosely?

Because of this looseness, when we try to transfer our knowledge to others, much of what we say is lost in translation. Often what's said isn't what is understood. Yet we still call that knowledge.

Words are the only tool with which I can communicate with you in this book. I'll try to use them as accurately as I can, but I know I'll fall short. This is why I often ask questions and depend on you to think through the concepts we discuss in depth. Only then will you find real knowledge. Taste your own mango. Don't take my word for it.

As a species, we built our full approach to knowledge on top of these flawed building blocks. A building is only as resilient as the materials it's made of, and sadly, our knowledge is as limited as our words are. When you put the depth, breadth, and limitations all together, you'll discover the only knowledge that seems to be true:

Remember!

→ **We truly don't know that much after all.**

Real Knowledge

When Ali was eleven, he bought a book called *The Ultimate Book of Useless Facts*. For weeks he would bring it on our weekend drives and read aloud the strangest things. "Every time you lick a stamp, you consume ⅒ of a calorie." "Most American car horns honk in the key of F." "Most toilets flush in E flat." All unsubstantiated, all useless lines of "knowledge"—yet they earned the right to be compiled in a book. He would laugh so hard his whole body would be shaking and he'd say "We humans are so silly!" Indeed we are, and I'm the first to admit it.

I was an addicted learner throughout my youth. I idolized knowledge. I was arrogant about what I knew and defended it fiercely—right up until I worked at Google. The first few months there shattered my Illusion of Knowledge. The novelty of the Internet presented me with so much that I didn't know despite my years of experience and forced me to reexamine what I thought worked. Walking the corridors of a place where everyone was much smarter than I shook my faith in the idea of "the one correct answer" that we all learn in school. Many members of our diverse team saw the world from so many different angles and often discussed topics in which several points of view were correct. The extreme data-driven approach to decision making, on the other hand, frequently exposed the invalidity of some perspectives. Sometimes ideas that were originally defended passionately turned out to be wrong. But the openness encouraged people to speak up and brought tremendous diversity to the conversation. Often a twenty-year-old would challenge the views of a more experienced vice president and turn out to be correct. After a year at Google I realized that I knew very little compared with all that is out there to know. I knew so little, in fact, that it felt as if I knew nothing at all. Luckily, my eagerness to learn overpowered my ego's desire to be right. This revelation gave me a deep sense of joy as I relinquished the endless struggle to defend my view and just enjoyed the journey of endless learning. That joy drives me, even as I write this book, to frequently pause and ask if what little I know matches your perception. It drives me to ask that you question its validity and find your own truth. If some of what you read turns out to be wrong, please forgive me. It's inherent in the nature of knowledge to occasionally be wrong. Reach out and inform me, then together we can learn a bit more.

If you have been shielded from the Illusion of Knowledge, then you are one of the lucky few. It took me years to learn to acknowledge that regardless of how passionately I believed what I know to be true,

I might still be wrong. There is always a chance that I missed out on an important detail, and there is always something more that I don't know. I'm not always right—that I know to be true.

The Russian Nobel Prize–winning physicist Lev Landau once said, "Cosmologists are often wrong but never in doubt." That is an admirable statement, especially from such a renowned scholar. A history of his field confirms the truth of it. In cosmology, we first assumed that the Earth was flat; when we acknowledged that the Earth was round, we were convinced it was the center of the universe around which all other heavenly objects revolved. Every step of the way, those who were wrong weren't at all in doubt.

Einstein, brilliant as he was, didn't arrogantly assume ultimate knowledge. He once said, "In theory, theory and practice are the same. In practice, they are not." When he made a huge mistake in attempting to "fix" his equations by inserting a constant to adjust for the effects of gravity, he eventually conceded and admitted he was wrong. Strangely, he was later found to be wrong *about being wrong* when scientists discovered that his arbitrary fix, the cosmological constant, was in fact one of the universe's most fundamental truths.

We shouldn't blame ourselves, however, for being protective about what we think we know. How could we go on doing what we need to do if we believe it is based on false assumptions? How could someone passionately engage in doing something if she believes it's wrong? Everyone, even wrongdoers, need to somehow find logic to justify what they do.

The more we know, however, the more we realize that we have glimpsed only a fraction of the truth. Confucius said:

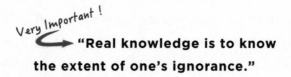

Very Important !

"Real knowledge is to know the extent of one's ignorance."

Be Wise to Be Happy

Knowledge is the illusion that keeps us from seeing the reality behind all the other illusions because it makes us think, *If we managed to get so far in life, then our knowledge can't be all wrong*. Indeed, you have gotten so far in life despite the six grand illusions and that eliminates the urge to debate their validity. But please be wise. Entertain the idea that what you've spent your entire life learning may not be entirely true.

Knowledge is in no way a prerequisite to happiness. Your default state before you had any knowledge was happiness. As a matter of fact, false knowledge is the underlying reason for most unhappiness. Our conviction that all we know is true leads us to use such knowledge as the input to our Happiness Equation. By the time we find out that what we know is actually false, the equation is already dysfunctional and the suffering has already set in.

If you examine the thought forms that cause you to be unhappy, you'll realize that they mostly stem from attachments to illusions and false beliefs. The concepts that have the deepest impact on us are the ones we believe most strongly to be true—when usually they're not.

Hanging on to false concepts is a bit like ostrich behavior: hiding your head in the sand, believing that you are safe while leaving yourself vulnerable to suffering. Not a clever strategy. So why do we do it? Because of ego.

The Illusion of Knowledge is strongly supported by the Illusion of Self, particularly the ego. We identify ourselves with our knowledge. We defend what we know and get offended when it's attacked. Since what we think is true is often different for different people, the attacks become frequent. It becomes a constant struggle to try to defend an ego. Undress. Leave your knowledge open to attacks. Be wise. Define yourself by openness to those who contradict what you "know." Be an explorer, a seeker of the truth, always ready to admit being wrong in order to continue the quest.

Please think about that for a minute. Think of a few times when something you believed to be true surprised you by turning out to be the furthest thing from the truth. You'll find you can remember quite a few examples. Please don't read on until you do. It's important to admit the extent of your knowledge to yourself before I throw in the closing point.

Explorers ready?

The Nudge

Our conviction of what's good and bad seriously complicates our approach to solving the Happiness Equation. When we looked at the Illusion of Self, we found that not acting as though you are the star of the movie helps you see that the distribution of good and bad evens out among the rest of the cast. There's more to understanding the concept of good and bad.

Of course, we always want the good things in life but seem to frequently get bad things. When the Happiness Equation is solved incorrectly, the world appears to have failed to meet our expectations. When that happens, you think of the current events as bad and you feel unhappy. Life, however, sometimes needs to give you a nudge in order to alter your path. It uses a bit of hardship to lead you to something good.

In 1990, a twenty-five-year-old Scottish woman named Joanne was on a train trip from Manchester to London when the train was delayed for four hours—an event we normally consider to be bad. During the delay, the idea for a story about a young boy attending a school of wizardry "came fully formed" into her mind. She started writing as soon as she arrived home. Two years later, however, only three chapters were written. And so life took charge and moved her to a job as a foreign-language teacher in Portugal, introduced her to a man who became her husband and the father of her daughter, but then took her through a painful separation. She had to go back to Scotland. By then the whole world seemed to be against her. Her marriage had failed, and she was jobless with a dependent child. The world wasn't against her. She was being *nudged*. The world was pushing her away from an ordinary life to greatness.

We know that because she went on to write the Harry Potter books as J. K. Rowling and thrill millions of readers around the world. Later she described that period of her life as liberating, allowing her to focus on writing. Life, with all its might, blocked every other path, leaving her only one, and she followed it. She made the most of her path, and two years later she finished her first manuscript. Her full series, in sixty-five languages, has sold more than 400 million copies, making her one of the top-selling authors in history.

Remember!

➜ **Sometimes when you stray off your track, life nudges you hard . . . and that's not bad!**

The Eraser Test

Life isn't selective when it comes to nudges. While it can nudge harder when there is greater good further down the path, it takes everyone on detours every now and then. Even you, I'm sure, have been taken on many such detours.

Here is a simple test to help you recognize when you're being nudged. It's the eraser test. Imagine a new technology that allows you to choose any past event you don't like and simply erase it as if it never

happened. Erase it from your memory and from the actual flow of time as well. The technology manages to pinpoint that exact event in the space-time continuum and uses a rubbery algorithm, written in just a few thousand lines of Python code, to wipe it clean. The technology also traces the effects of the event across the trail of time and automatically erases all of its consequences all the way up to the present moment.

Since you're one of my esteemed readers, I'm going to let you test-drive this new technology. You can choose to erase anything. You could choose to erase a boring lecture along with everything you heard in it; by doing so you will erase every interaction with anyone you met at the time, every call you made to those people afterward, every bit of information that stuck with you—everything. The eraser will also give you back the time spent there, so it will make it seem as if you arrived home an hour earlier through a different route, listened to a different program on the radio, and so on. Your path in life will be altered to correspond with one that isn't affected in any way by the event you elected to erase. Hopefully this new path will be one you're happier with, but no guarantees are offered as part of testing the eraser.

Go ahead now, pick an event you want to erase and consider all the other twists and turns of life you'll erase with it.

Now ask yourself, if this technology actually existed, how many events would you elect to erase? Ask yourself how many events you would rather keep even though at the time they happened you thought of them as bad.

Almost all of those to whom I've presented this hypothetical test agree. While a few managed to find a handful of events they deeply regretted and wanted to erase, most decided to keep some of their toughest experiences unchanged. Knowing that erasing an event erases the trail it created, the majority of people I interviewed chose to keep the nudge, grateful for the path onto which it took them. Some even said that looking at those experiences as nudges makes it clear that they might have been the best things that ever happened to them—though this was not always easy to see.

In retrospect, I can clearly see how my own path in life was helped by a series of spectacular nudges. Yet I'd use the eraser once if I could. I admit that I'd still truly love to erase Ali's departure. There's nothing I would love more than just one more hug. I understand that his departure was the biggest nudge of my life, directing me to write this book and do more good for others, but even as I control my thoughts, my heart will always miss him. I guess time—a lot of time—will heal what positive thought cannot.

Arabic folk culture tells the story of a wise old man whose son went to the well one day when the heat was at its most extreme. To his surprise, he found a beautiful, tame, black Arabian horse there. Everyone in the village envied the young man, and he started to win all of the races with his new horse. The villagers told the old man, "Your boy is blessed with good fortune." To which the old man responded, "You never know for sure." A week later the boy fell off the horse and broke his legs, so they rushed to the old man and said, "Your boy's good luck has turned bad!" To which he responded, "You never really know for sure." A week later, a rival village suddenly attacked, all the able young men were drafted, and many were killed. The man's son was spared.

Remember !
↪ **You truly never know for sure.**

Now, please be honest: How many of the worst things that you've faced turned out, in time, to be the best things that ever happened to you? How many made you the person you are today? How many helped you meet someone you loved or taught you something you needed to know? I know that many of those experiences were harsh, and some still hurt, but how many were all bad, so bad that you'd be willing to erase them?

When you realize that every seemingly bad event nudged you onto the path of many good events, you'll reset your definitions of good and bad. The new definitions will help you make amends with the Happiness Equation. You'll realize that your expectations are sometimes rushed and that life can surprise you by eventually coming around to work in your favor. It so often has in the past. Why would it change now?

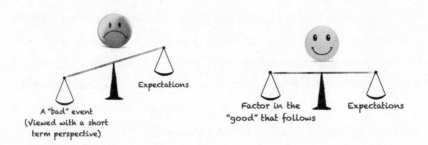

A "bad" event (Viewed with a short term perspective) — Expectations

Factor in the "good" that follows — Expectations

Every moment of your life is neither all good nor all bad. When you clear your thoughts and see beyond the Illusion of Knowledge, you will realize that what Shakespeare wisely said is true:

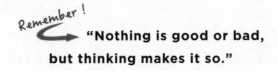

Remember!

"Nothing is good or bad, but thinking makes it so."

Does Anybody Know What Time It Is?

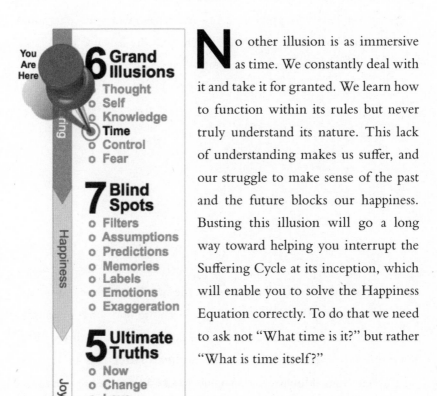

You
Are
Here

6 Grand Illusions
Thought
o Self
o Knowledge
Time
o Control
o Fear

7 Blind Spots
o Filters
o Assumptions
o Predictions
o Memories
o Labels
o Emotions
o Exaggeration

5 Ultimate Truths
o Now
o Change
o Love
o Death
o Design

ring

Happiness

Joy

No other illusion is as immersive as time. We constantly deal with it and take it for granted. We learn how to function within its rules but never truly understand its nature. This lack of understanding makes us suffer, and our struggle to make sense of the past and the future blocks our happiness. Busting this illusion will go a long way toward helping you interrupt the Suffering Cycle at its inception, which will enable you to solve the Happiness Equation correctly. To do that we need to ask not "What time is it?" but rather "What is time itself?"

A Timeless Experiment

Let's pretend you have volunteered for a mysterious research mission. All you've been told is that you will be confined to a small spherical capsule and propelled along a track between two cities. No one will say how much time it will take to get to your destination, but you will be stopping at a number of stations along the way; the research team tells you to expect to stop every ten minutes or so. There's nothing in the capsule but your seat: no windows, no dashboard, no entertainment. Just a smooth, blank, brushed-aluminum interior.

As you're being strapped in, a researcher removes your watch and takes your cell phone. You start to ask why, but she cuts you off. "Time to go," she says as the door slides shut. "See you on the other side!"

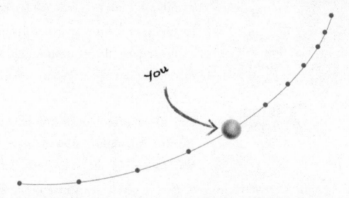

With nothing else to do, you count the number of times your capsule slows down and accelerates. By the time the door of your capsule reopens, you've felt the ease and push of G-forces a dozen times. With ten minutes between stations, that means the trip took two hours. That feels about right. However, after you've exited the capsule you're informed that the researchers who checked you in were mistaken: the travel time between stations was five minutes, not ten. What you had thought was a two-hour trip took only one. But the first researchers seemed quite certain about the information they gave you. Which team of scientists should you believe?

You rack your brain, but no answer comes to mind. Quite simply, there isn't one. If you had been in that capsule long enough, you'd eventually have felt hungry, thirsty, and tired. But those are markers of *biological* time. You still wouldn't have known exactly how much time had elapsed according to the clock, because *mechanical* time is purely a human construct.

Ancient civilizations used the arc of the sun through the sky to measure time of day, the lunar cycle to calculate months, and the seasons to gauge the passage of years. We developed sundials, obelisks, water clocks, hourglasses, and, eventually, mechanical, digital, and atomic timepieces. Today we measure time so precisely that we've come to think it exists independently of the arbitrary measurements we've established. But it doesn't. Even atomic clocks, which are responsible for synchronizing time for much of our technology, including electric power grids, GPS, and the watch on your smartphone, need to be tweaked in order to stay synched with nature. Over thousands of years, day would eventually become night if we didn't insert what's known as a "leap second" into Coordinated Universal Time every four years to keep this from happening.

All the devices we've invented so far are working away to tell us what time it is, but no one really knows what we're actually measuring. We're just using mechanical movements to track the passage of time because that's an approach our brains can latch onto. But the nature of time itself is still illusory. Your experience in that capsule trip was a clear indication of just how arbitrary our measurements can be. Now let's take another quick ride, but this time we're going to eliminate the station stops altogether. It'll be one smooth trip with a jolt of acceleration at the beginning and no perceptible movement

after that, ending even-
tually whenever it comes
to a stop. On this second
journey you'll have abso-
lutely no way to tell how
much time has elapsed
when that door reopens.

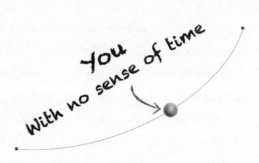

This time your journey, with no mechanical movement to measure,
will be "timeless."

Who's Your Master?

Over the past few centuries, time evolved to become the taskmaster of
modern culture. Squeezing more tasks into the limited hours of a working day has meant greater production and higher profits since the start
of the Industrial Revolution, and where profits have gone, society has
been sure to follow. We live by our watches. When the alarm goes off,
we wake up and start the day. When the bell rings, we move from one
classroom to the next. At the appointed hour we move from meeting to
meeting, from a phone conference to an errand to a dinner with friends.
We fall asleep having double-checked the alarm, knowing we'll respond
to its command to rise in the morning. Time is woven into every plan
and every action. We depend on time, and we fret under its pressure.

This has certainly been true for me. Time has always been one
of my biggest challenges. That's partly my own doing, since I like to
set ambitious goals for my workday. And while I could just meet my
clients online, I prefer to see them face to face, so I put up with flight
delays, getting stuck in traffic, and rushing around to get to meetings
on time. In the past, time caused me a lot of stress and a ton of unhappiness. So I set out to improve my relationship with it. First, I learned tips

and tricks for managing tasks and calendars in order to become more efficient. But that was just scratching the surface. I needed a deeper understanding of what it was I was trying to manage, and that's when things got interesting.

The first key to the Illusion of Time that I became fascinated by was that it's so subjective on a personal level. Its passage feels different depending on the situation you're in. Have you ever had the experience of being in a serious car accident? I flipped my car from a bridge once. I hope this never happens to you, but what I observed was extremely interesting. Time slowed down dramatically as my vehicle approached the point of impact. I can't resist calling this slow-Mo ☺. Soldiers in combat report the same phenomenon. And I'm sure you've noticed that a boring two-hour lecture goes by much more slowly than an evening with good friends.

This feeling of relativity in the way we experience time reminded me of one of my great passions as a young boy: Einstein's theory of general relativity. So I went back to the books to see what science could teach me.

The Science of Time

Our scientific understanding of time took a huge leap forward at the turn of the twentieth century. Prior to that, classic Newtonian physics taught that time always had an absolute value. Mathematicians plugged it confidently into their equations until Einstein's theoretical breakthroughs were published in 1905. His theories of special relativity and, later, general relativity represented a dramatic shift in our relationship to time. Einstein asserted that time and space are not two separate things, that they combine to create a four-dimensional structure he called space-time.

Einstein further explained that the pull of gravity actually slows time down. So if you were an astronaut on a long interstellar trip and your spacecraft passed close to a black hole (where the gravitational force is massive), time would slow down significantly. When you got back to Earth you might have aged several years, but your spouse and your friends would have already lived into old age. We can observe this effect in a much smaller way right here on Earth. If you lived in Dubai on the top floor of Burj Khalifa, the world's highest tower, time would pass slightly faster for you than it would for someone living on the ground floor, just because gravity affects each of you differently. While a variance like this is too small for the human body to detect, it's measurable with today's technology.

It gets even more bizarre. The math indicates that in space-time, past, present, and future are all part of an integrated four-dimensional structure *in which all of space and all of time exist perpetually.*

Imagine space-time as a loaf of bread, where every slice represents everything that is happening anywhere in the universe in a specific instant. We humans can move in different directions in space, but we experience the dimension of time only slice by slice as we move through it. If we had the ability to perceive time as we do space, we would be able to hop back and forth through time as if we were

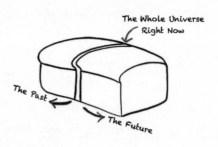

getting on and off a train at any station we please. You hop onto the train in 2015 and go visit the Jurassic period—not that I recommend it—which is there for you to enjoy in that epic real-life movie called space-time. Man, that stuff still makes my head hurt after all those years of studying it.

Does this sound hard to believe? It should be. It is hard to believe. But the math speaks clearly. And if time really is this strange and exotic, then why do we feel we know it so instinctively? The space-time bread loaf feels nothing like the time we know day to day, the kind that gives us so much stress. In fact, time gets even weirder and less familiar the more we look into it.

The reason we're usually limited to our own particular slice of space-time is due to a phenomenon in physics called "the arrow of time." It's the reason we have the freedom to move anywhere in the three dimensions of space but can exist only in the "now" slice of space-time. This is a tricky concept. It will be easier to understand if we move from loaves to trains.

Imagine that the whole universe is squeezed into one train: every galaxy, star, and planet and every grain of sand and living being. This train sets off on a journey, not from one city to the next but rather on a journey through time. As a passenger on that train, you can move anywhere you choose, but you can't change the train's direction or speed, which is restricted to a track that is the arrow of time. You just go for the ride with no control over its position or orientation, hopping from a slice of "now" to the next along the time dimension.

Remember!

Time isn't moving; you're the one who is moving through time.

So we are always positioned to experience the current slice of now, but even that is relative. According to Einstein, *the speed at which you move affects your experience of time.* On his journey back to Earth, that astronaut I mentioned earlier would be moving at astronomical speeds that would lead him

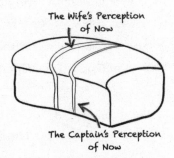

The Wife's Perception of Now

The Captain's Perception of Now

to slice the space-time loaf at an angle. As a result he would have a very different perception of "now" than his wife back home on Earth. Like the rest of us she might experience a time slice that included events in different parts of the world, all of which were occurring at the same time. She could be feeding her child while watching a TV newscast showing images of war in Syria, while listening to a friend cheer in the other room as the New England Patriots win their fifth Super Bowl. These events are happening simultaneously as perceived here on Earth inside her time slice. But her husband's astronomical speed would lead him to see the different points in space *at very different times*. As his speed warps space-time, his perception of closer points on his travel path reach him quicker than the farthest points, so he ends up slicing the loaf at an angle. As a result, his time slice might include the destruction of the Berlin Wall in 1989, his mother's death from cancer in 2016 in the United States, and the discovery of a cure for that cancer ten years later in Japan! All those events will seem to him as if they're happening at the same time because they're happening in a different "place" on his slice of space-time. Ouch, my head hurts again.

The reason I'm going on about the complex scientific aspects of time is because **the more you know about time, the more you'll appreciate that in reality it's nothing like we think it is**. Time as we generally accept it is an immersive illusion that has very little to do with what it really is or how it behaves. The time we think we know doesn't exist (as you saw in the capsule experiment). Nothing about the way we individually experience time is absolute.

Remember!

Time changes for everyone.

As time changes and morphs from one person to another and from one situation to the next, one can't help but conclude that it isn't real. Time completely fails the test of permanence. Time is an illusion because, in an interesting way:

Remember!

⤷ **Time completely fails the test of time!**

A Cat's View of Time

Theory aside, we're all constantly stressed by time. You've got a major deadline for a project at work, and the stress is making you grouchy. Or you're upset that you're single because your biological clock is ticking and you want to have a family. Or you're cursing at traffic because your daughter's waiting for you at an afterschool program and you're running late. So why isn't your dog or cat losing sleep over any of this? No other known life-form responds to the ticking of a clock or counts the days, months, or years. As you've noticed, animals simply respond to events. Hungry? Let's eat. Dark? Let's sleep. It's a way of living that has more than a little appeal!

Even within our own species, human cultures differ widely in their views of time. I, for example, grew up in a culture that was more events-based than clock-based. So I find it odd that countries in the West place such a premium on meetings that begin and end "on time." I've been in meetings in the United States where everyone starts to put away their papers and get ready to leave as the scheduled conclusion approaches, *even if a few more minutes might lead to a big breakthrough.*

Delaying the next meeting in order to focus on the big opportunity at hand happens rarely, if ever.

By contrast, in the Middle East and Latin America events tend to have a loosely defined start time and continue for as long as they seem worthwhile. If a meeting's going well, it'll get as much time as it needs. Similarly, when planning a social gathering friends might agree to meet after work, but some will show up at 7:00, some at 8:00, and some at 11:00, and no one will be offended or stressed because everyone will enjoy the company of whoever is at the event for as long as they're there. It's a bit like a cat's attitude toward time.

You might be surprised to learn that events-based cultures are more common globally than clock-based cultures. While they may seem laid back in comparison to their efficiency-focused Western counterparts, businesspeople in many parts of Latin America, the Middle East, southern Europe, the Indian subcontinent, and Africa are extremely adept at establishing social connections and working together. They're solving for their own definition of success. But as they do, they're also solving for happy.

While I surely don't advocate being late, slack, or lazy, I would ask you to consider the merits of being the master of the task instead of being the slave of the clock.

Even more at odds with the West are those Asian cultures we call "timeless," of which Buddhism is the best known. Clocks will tick and events will come and go, but a Buddhist stays fully focused in the present moment. This state of timelessness is fundamental to reaching Nirvana. The ability to live entirely in the present moment offers the peacefulness of living in an eternal paradise, especially when you realize that while eternity is commonly understood to be a very long time, it really is the absence of time. It is timelessness.

Remember !

Time is experienced very differently across human cultures. There may be a better way to deal with time than what we're used to.

Time as we understand it plays a big role in creating—and perpetuating—unhappiness. When it comes to feeding the Suffering Cycle, time affects us at a profound level. To escape its shackles, let's get back inside the brain and into the thoughts it generates in relation to time.

Past and Future

In chapter 3 we learned about the voice in your head and became aware that you're more than just the steady stream of thoughts being generated by your brain. This is important to solving for happiness because that voice is often the messenger of thoughts that lead to pain and sorrow. When you scrutinize the thoughts themselves, you'll notice that very few of them have anything to do with the present moment. They're almost always entangled in the past or busy predicting the future. That's particularly true of unhappy thoughts. Almost all emotions anchored in the present moment, though, are happy (note that physical pain is not an emotion). Interesting, isn't it?

To test this out on myself, I sat down and sorted out all the emotional states I felt throughout my normal life. Then I assigned them a tense (past, present, future) and a valence (negative or positive). Here what I found:

	Negative			Positive	
	Active ←		Passive	→ Active	
Past	Grief, Embarrassment, Shock, Anger, Annoyance, Contempt, Disgust & Irritation	Hate, Despair, Envy, Guilt & Shame	Disappointment, Hurt & Sadness	Satisfaction, Pride & Appreciation	
Present			Boredom	Calmness, Affection, Empathy, Friendliness, Love, Courage, Pride, Satisfaction, Trust, Contentment, Relaxation & Relief	Amusement, Delight, Elation, Excitement, Interest, & Surprise
Future	Anxiety, Fear, Helplessness, Powerlessness, Worry, Stress & Tension	Doubt & Frustration	Pessimism	Optimism & Hope	Courage

Wait! Wait! Don't skip over the chart. Take the time to study it in detail.

While all emotions are felt in the present, they tend to have an anchor in the past or future. Regret, for example, is felt in the present moment but focuses on something that has already passed. Please take a moment to examine the Negative-Positive Chart. Read the lists of feelings linked to the past and future, then compare them with feelings linked to the present. There's a significant difference, don't you think?

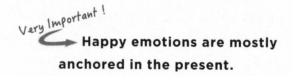

Very Important!

→ Happy emotions are mostly anchored in the present.

Now ask yourself this question: Have you ever experienced *anything* that happened outside the present moment? I know at first that sounds silly, but take some *time* to stop and think about it before firing off a quick answer.

The past that we consider such an important aspect of ourselves is really nothing more than a record of moments we call memories. It's just a collection of thoughts in your brain, and a famously unreliable collection at that. Despite the temptation to see the past as real, it is not. The only time that ever really exists is the moment you experience as *now*, and once that moment is replaced by the next, we call it the past. This applies to this moment as you read these words. Oh, I mean this one. No, this one. Once the present moment passes by (and it doesn't take long!), it no longer exists. It's gone. Forever.

Nothing ever happens in the future either. How could it? The future hasn't occurred yet, and it only ever will if all its infinite possibilities collapse into a moment that takes place in an instant of now. We can absolutely say, then, that when your thoughts and feelings are caught up in the future, you're just imagining things! Moreover, you have absolutely no way to guarantee that out of all the endless possibilities for how things could turn out, the one you're imagining will be the one that actually takes place. What are the odds? As someone who loves math, I can tell you that they're not good!

Unfortunately for happiness, your brain is sold on the idea that the next moment is more important than the one we're in. On the other hand, the moment that already passed by is more *familiar*—and therefore perhaps more comfortable—than this one right now. These biases of the brain are what move us all too easily into a state of confusion, ruminating on the past or bracing for an imagined future while neglecting to pause and live in the present, even though the present is all there really is.

Very Important !

When we're focused on the past or the future, we're living in our thoughts, and not in reality.

The impact of living in thoughts is beyond profound. Have you ever noticed how quickly your life has gone by? Doesn't it sometimes feel like the past twenty years have just evaporated without your even noticing? There's a good reason for this.

If you're not in the here and now, then you're simply in your head. There's nowhere else to be. If the past twenty years feel more like a week, it might be because you spent only a week of that time truly experiencing life, being fully present. For the remaining ten and a half million minutes you were just wandering around in your head. What a waste!

Am I saying that all your memories are worthless? Not at all. There are wonderful memories based on times when you were in the now. The moments I'm talking about are the ones when you were in your head worrying about the past or the future. You don't remember those thoughts because they were not memorable. They're wasted because they *could* have been in the now, making memories!

When I think about the way we waste our lives, I can't help but remember the lyrics of Pink Floyd's song appropriately titled "Time." Specifically this line: "And then one day you find ten years have got behind you. / No one told you when to run, you missed the starting gun."

Your Brain Is Addicted to Time

If past and future thoughts make you unhappy and can lead you into wasting big parts of your life, then why is it so difficult to focus on the now? What is the defect in our design that makes it so hard to escape our preoccupation with the past and the future, even though it increases our suffering? It's this: time is the ground from which thought itself is grown.

To get a better sense of this, let's try my favorite exercise. I call it the Full Awareness Test. Sit down, take a deep breath, relax, and close your eyes. Keep them closed for a minute or so. The time doesn't have to be exact; you're just clearing your visual palate. Now open your eyes for a few seconds and look around. Don't do anything more than observe whatever is in your immediate environment. Then close your eyes again.

Now, with your eyes still closed, silently describe to yourself what you saw. This isn't a memory test; we're just looking for a description of whatever it is that you've observed. Go into detail about as much as you can remember, but be as factual as possible. Be careful not to let your thoughts intrude and interpret what you've seen. Stick with statements like this: *I look out my apartment window at night and see water extending to a low horizon of palm trees and two-story buildings interspersed with tall, gleaming skyscrapers. Daylight is still fading from the horizon, where I can make out thin strands of clouds. Higher up, the sky is much darker and pricked with stars.*

Are you noticing that it might take you several minutes to describe what you observed in just a few seconds? The act of observation requires only simple awareness, but describing it introduces much lengthier thought processes. Still, because you're limiting those thoughts to describing what you've just seen, they're always in the present tense. They have no past or future timestamps, and as a result they're smoother and calmer than they would otherwise be.

You see? Connecting to the present doesn't require much brainpower at all. If your brain were limited to only describing whatever is going on around you in this moment, it wouldn't have much to say. The voice in your head might sound something like this: *Blue sofa ahead. Two light sources. A floor lamp and a candle. Candle flame flickering in the breeze. The smell of freshly baked bread.* Not much of a conversation, is it? That's because there are no pros and cons. There's no drama until we throw in the past and the future.

I frequently turn to this exercise to get myself anchored in the present moment. It calms me and reminds me of two important features of awareness: we don't need thoughts to be aware, just *presence*; and even when we do introduce thoughts, by focusing them on the present moment we become much less stressed. Giving your brain the task of simply describing its surroundings keeps those thoughts quieter, smoother, and easier than the incessant stream of thoughts that takes you beyond the here and now.

Try this exercise again. You don't need to close your eyes once you get the hang of it. Notice the cup of coffee on your table but resist the temptation to label it good or bad, hot or cold, or to wonder whether it'll end up leaving a ring on your wood furniture. Just limit your thoughts to what you see in this moment: *A white ceramic cup half-filled with black coffee on a bleached pine table.*

When you tune in to the present moment you accept it as it is. You don't compare it or judge it, and you don't wonder how it could be different in the future or better or worse than it was in the past. You are 100 percent in harmony with the Happiness Equation. Events totally match your expectations when you observe the world as it is. How peaceful! ☺

Most of our thoughts, however, do come with a timestamp. They're based in the past or in the future, which makes them far more likely to lead to unhappiness. To make a judgment you need to compare a current observation to one you've made in the past. To be anxious you need to think about the future and anticipate that it'll be worse than the present. To be bored you need to long for a state other than what's happening in the present. To be ashamed you need to re-create a moment that no longer exists. To be unhappy you need to focus on what you want that you don't yet have. With the exception of pain, no one ever suffered from what was going on in the present moment. Think about that for a minute. It's true. Even for you.

Without time, the Suffering Cycle is interrupted at its inception. Remove time, and the original thought fails to arise. Every stressful or unhappy thought exists outside of the here and now, while every observation of the here and now eases you into a peaceful place. Time and mind are inseparable. When you remove the timestamps from your thoughts, there will be nothing unhappy left to think about.

Very Important !

If you want to be happy, live in the here and now.

Use Time: Don't Let It Use You

At this stage of the conversation, I've found that a lot of people start to defend the importance of time. They ask, "How could I even function if I don't plan for the future? Who would I be without my past? The whole world runs according to the clock, so how do you expect me to start ignoring time altogether? What about the fact that I have to get to work by nine o'clock tomorrow morning?"

These are good questions!

For the purposes of this conversation, let's say that there are two types of time: clock time, which has some practical uses, and brain time, which doesn't. Clock time has to do with a specific event positioned at specific coordinates in mechanical time. Thoughts associated with clock time are practical, actionable thoughts, such as *I wonder how long it will take me to get to my doctor's*  *appointment.* They lead to precise answers, such as *That usually takes twenty-five minutes, but I'd better add another fifteen to account for rush-hour traffic.* They don't get complicated or layered with emotion, and they don't linger. Planning to be on time for a date or to pick up your kids from school and setting aside time for the things that are important to you are examples of clock time. They're logistical and they're benign. They help us meet our obligations. They keep us on time.

Brain time, however, tends to get caught up in thoughts about the past and the future. It gets lost in endless—and unlikely—scenarios of how an event in the future might turn out. Brain time obsesses over past events that didn't turn out the way you hoped they would. Thoughts in brain time tend to jump from one to the next. They don't lead to a specific action, and like dreams, they're formless. When you get hung up in brain time you might look up to discover that, according to clock time, the hourglass of your life is much closer to empty than you'd ever imagined.

As long as your thoughts are describing events in the present moment or are oriented toward clock time as a way of meeting your practical needs, you're fine. So keep an eye on your thoughts to see if they're time-stamped. If they are, then they've veered off from the present and probably aren't solving any practical problem. Your thoughts wandered into the endless confusion of past and future. Stay in the moment, and you've succeeded in stepping out of the most pervasive of the six grand illusions, the Illusion of Time.

But What if Now Isn't Happy?

When I talk to people about time and happiness, one comment I hear a lot is this: "But for me, being in the present moment doesn't work.

I'm unhappy right here and right now." When we sit down to explore why things aren't happy right now, however, the conversation always goes something like this: "I'm ashamed that I wasted last year partying and didn't pay enough attention to my schoolwork. My grades were terrible, and now everyone thinks I'm stupid." I respond by pointing out that this painful thought might be taking place right now, but it's anchored in the past. Nothing is going to change what happened last year. That was then. We're in the present moment, and all you can do in this moment is focus on your studies and work hard so that you'll do well right *now*.

Someone else will tell me, "I'm unhappy right now because all the guys I meet are jerks. I'm sure I'll never find the right one, and if I don't I'm going to spend the rest of my life alone." My response? Your unhappiness is tied to thoughts anchored in the future. How do you know that the love of your life isn't sitting alone at a café just around the corner? Or that getting married in a hurry won't make you a lot more miserable than you are right now? Enjoy yourself in this moment, and you'll find lots of people who'd like to join you for the ride.

Another person will say, "But I didn't get that promotion I worked so hard for. I can't believe that I'll have to spend the rest of my life in this dead-end job!" This thought, I point out, is anchored both in the past and in the future—but not at all in the now. Then I ask, "How can you know that for sure? The next job that comes along might be much better suited to you. Yes, it feels disappointing not to get the job you were hoping for, but the hard work you've done is already in the past. Thanks to that dedication you are more capable, more competent, and more seasoned right now."

Every time you examine your thoughts you'll notice that whatever you're upset about is rooted in a past you cannot change or a future that may turn out to be completely different from what you expect. You may as well let the past or the future go and do your best at whatever you're doing now. This is the moment, the only one you can count on. Live in it fully, and the rest will take care of itself.

It can sometimes be hard to stick to this approach. For example, a friend once told me, "For me, this present moment you talk so much about is a nightmare. I just got off the phone with the doctor, and the biopsy shows I've got cancer, stage four. The doctor doesn't recommend surgery, and he's giving me between six and eighteen months to live." To which I responded, "That's rough news. I'm very, very sorry to hear it. I wish you didn't have to go through this. In the meantime, maybe you don't have to add suffering on top of a difficult diagnosis. While we work to find you the best treatment, I'd like you to remember that *right now* you're alive, so savor every sweet second that you're on this planet with your friends and family. This is truly all you can control. I know it doesn't feel that way, but don't forget that you're no different from the rest of us. Anyone and everyone you know could leave this earth within the next eighteen months, or the next eighteen days. The only difference is that the rest of us aren't thinking about it, so we're not suffering. As far as the future goes, be optimistic, but live *now*. Take away that thought of what might happen in the future, and it won't make you suffer."

Very Important !

In this very moment there is absolutely nothing wrong at all.

Live Here and Now

The title of this chapter poses the question "Does anybody know what time it is?" That was a trick question. Regardless of where you are in the world as you're reading this book, *the time is now*. There will never be any other time. Any other interpretation of time is just a detour into illusion.

Before I end this chapter, I'm going to ask you to step back into that capsule for one more trip. This time the researcher proudly informs you that the technology now moves the capsule *instantaneously* between stations. It takes no time at all to get to the other end. "We've also added lots to experience at every station," she says, "which is why some volunteers have complained that the journey passes by unnoticed while they would've preferred to take their time and enjoy it. So we added a new feature, a button that you can press whenever you want to move forward to the next station. If you don't push the button, the capsule will automatically advance from each station at midnight every day. You can press the button seventy-five times at your own convenience to get to the other side or experience the full seventy-five-day trip. It is your choice." She then says, "Or was it seventy-five years? I can't remember. It doesn't matter, it will pass by in a flash either way."

"That's an easy choice," you say. "Let's get on with it. See you on the other side." As she closes the door she says, "Oh, I forgot to mention, I won't be able to meet you there. When you reach the other end, you die. This ride is all you have." She pushes the start button, closes the door, and sends you on your way.

Now that you're in control (sort of), will you press the button quickly and get it done with, or will you spend every day experiencing each station to the fullest? Will you spend the time in each station thinking about Day 75? Will you spend it regretting the days that passed? Or will you spend every day experiencing that day and everything it has to offer?

Make up your mind.

Very Important !

Life is now and now is amazing.

Chapter Seven

Houston, We Have a Problem

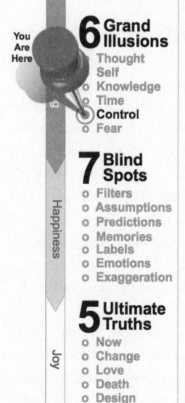

You Are Here

6 Grand Illusions
Thought
Self
o Knowledge
o Time
Control
o Fear

7 Blind Spots
o Filters
o Assumptions
o Predictions
o Memories
o Labels
o Emotions
o Exaggeration

5 Ultimate Truths
o Now
o Change
o Love
o Death
o Design

Happiness

Joy

Y ou put the pen down with a big smile on your face. *That was a lot of work*, you think, *but it was worth every bit of effort.* You take a sip of your coffee, sit back in your seat, and read your notes one more time. *This percentage of my monthly income goes to my retirement plan, and that percentage goes into the savings account. A standing order will pay my credit card and mortgage. I got the car insurance, life insurance, health insurance, disability insurance, homeowner's insurance, credit card fraud insurance, and on top of all of those, the umbrella insurance.*

Have I missed anything? Nope. Looks complete, you say loud and proud, even though there's no one around to hear you.

You lean forward, grab your calculator, and go over the numbers one more time.

Click, click, check. Everything's in order. You throw the note pad on the coffee table with a confident flick of the wrist and stretch back with your hands behind your head. *Well done! You have it all under control.*

Those are the best moments ever, aren't they? When we feel we've truly done all our homework, thought through every possible scenario, followed expert advice, and planned a clear path forward. It gives us peace of mind that everything is under control and that we'll be just fine.

The Truth about Control

Many people I know, all over the globe, were in exactly this situation back in 2008. They thought they had it all figured out—right up until the U.S. housing bubble burst, triggering the largest economic crisis the world has faced since the Great Depression. The market crash took away most of the equity in their houses; the companies they worked for went belly up; some failed to pay their debts, and some even saw their homes foreclosed on. Over the span of a few months many went from "All figured out!" to "What just happened!?" Some recovered, and some are still suffering, but everyone learned that things can, and frequently will, go wrong. Very wrong.

Our need for security and control is instinctive. In other species, survival is a matter of running when the tiger shows up, but we humans carry the burden of being a lot more sophisticated. We can forecast risk and plan our escape route long before that tiger is even born. We can scan the terrain and identify every possible threat, including those that are wildly hypothetical. We can take preventative measures, erect fences, and add surveillance cameras. Furthermore, we can extend our plans to include those we love because we care for them—and because their safety is part of our emotional safety. This very human set of survival skills is partly why we're still here while so many other species are not. We're able to take control—or at least believe that we're in control—while the best other beings can do is to react appropriately when the trouble starts.

Since the beginning of the Industrial Age, humanity has taken that control to a whole new level. Laying a railway track, erecting a high-rise building, and mass-producing an iPhone require spectacularly intricate planning and control. Call centers where every word is scripted, delivery services with real-time tracking—the limit to how far we can go to eliminate uncertainty keeps getting pushed further and further out. Our ability to stay on point in the simulated and hypercontrolled environment called work makes us believe that we can meticulously control our personal lives as well. And I'm no exception.

Although life has given me more than I need and assured me a future of financial independence, I still plan meticulously. I've got my career mapped out to the tiniest detail five years ahead. I plan my investments, savings, and where I'll live, plans that naturally extend to cover my family as well. I bought properties to ensure our prosperity, planned the kids' education, and invested in insurance and savings plans so that my loved ones will have what they need even when I'm gone.

I had pages of comprehensive plans and then, well, you know what happened. Four days into our (well-planned) summer vacation, Ali was admitted to the "wrong" hospital, where an error measured in millimeters led to his departure. How about that for control?

This tragic, surprising event wasn't part of any plan. We say we can't plan for such a dramatic turn of events because they're so un-expected, but is that really true? How often do those kinds of events happen? All the time!

I know you might not like hearing it, but in the United States alone, medical errors are the third leading cause of death, with various estimates of the loss of life to be somewhere between a quarter and half a million deaths per year. In countries where malpractice litigation isn't as advanced, those numbers multiply to millions. Other human errors, such as driver error and violence, take the lives of millions more. Al-though unexpected death is all around us, we choose to think of it as exceedingly unlikely.

Similarly, we choose to ignore most other disruptive events that occur hundreds, thousands, and millions of times every day. Natu-ral disasters, economic crises, victimization by fraud, bankruptcies—life-changing, plan-altering events take place everywhere all the time. I call these events *left turns* because they point us down a road we weren't expecting to take. And our path through life seems to turn left way too often.

Swans and Butterflies

In his *New York Times* bestseller, *The Black Swan: The Impact of the Highly Improbable*, Nassim Nicholas Taleb demonstrates that rare and improbable events occur much more often than we dare to think. His examples include the outbreak of World War I, the 9/11 attacks, and the rise of the Internet. The repercussions of these unanticipated "black swans" touched every single life on the planet.[1]

Consider for a moment how many similar events have happened in your lifetime and how many personal black swans have shaped your own life.

Taleb argues that our blindness with respect to randomness, particularly to large deviations, extends much further than our conscious awareness can even comprehend. This dovetails with what meteorologist Edward Lorenz called "the Butterfly Effect," the ability of seemingly minor and unrelated events to cause major changes. Lorenz ran a series of weather models in which, after inputting the initial conditions, he added tiny changes in the wind speed. Even though these changes were almost imperceptible—he compared them to the turbulence created by a butterfly flapping its wings—the ultimate outcome changed significantly, leading to the speculation that the flapping of the wings of a butterfly in Brazil can cause a hurricane in Florida.[2] Trillions of butterfly effects are buffeting us every minute. They alter our paths more than we can imagine.

To consider Ali's life as an example, the black swan was the medical error, but many butterfly effects also led to the tragedy of losing him, including the proximity of our home to that specific hospital, the repetition of his easily treatable belly pains, and the germ that might have started the inflammation of his appendix. All these occurred months or years earlier. Could I have controlled or planned them all? No. Control is an illusion.

Very Important !

→ **Between black swans and butterfly effects,
nothing is under your control.**

The Span of Your Control

Before we jump into deeper water, I should highlight that it's not my intention here to depress you. As any successful businessman will tell you, success (which in our case is happiness) doesn't come from ignoring unpleasant realities. It comes from realism and objectivity in understanding life with all of its imperfections. Happiness comes from our ability to navigate such reality based on facts, not illusions.

Acknowledging our limited control shouldn't cause us to despair. Addressed head-on, it should lead us to a realistic path to happiness. It all starts with understanding the true nature of our control. We think we are in control of everything—our money, friends, and career. But, honestly, how much control do you really have over those things you're hanging on to? Take any example, say, your money. Is your money really under your full control? "Sure," you say, "it's my hard-earned money. I can do anything I want with it. I can choose to spend it, give to charity, invest it, or save it."

But can you really? What if your bank goes bankrupt? It's happened before. What if taxes increase? Have you considered how inflation is taking away from your money, your purchasing power, while you can do nothing about it?

Neither is your career fully under your control. Your company could go out of business or might decide to lay off the workforce. Nor are your possessions, friends, or health. We all lose things and people we love, and we all get sick sometimes. Which has to leave you thinking: is there anything ever under our total control?

Yes, two things are: **your *actions* and your *attitude*.**

Your Actions

As an engineer, a senior executive, and a businessman, I'm the worst when it comes to control. For years I attempted to assert full control on every aspect of my life. At work, I wanted everyone, every system, and every data point in my organization to fully match my expectations. In my personal life, I tried to control my wife, the progress of my kids, and even the number of full loads of laundry that ensured the optimum use of water and electricity at home.

But no matter how hard I tried, events in the real world defied me. So what did I do? I tried

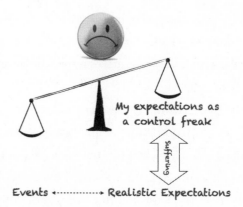

harder still. I was in a state of constant suffering, and it took me years of rejection, anger, and frustration to see the light and accept the truth: *I wasn't in control.* When I realized that, I felt a ton of weight removed from my shoulders. My actions remained committed, but my attachment to outcomes completely vanished.

My first breakthrough came when a friend taught me about the Hindu concept of detachment, when you strive to achieve your goals knowing that the results are impossible to predict. When something unexpected happens, the detachment concept tells us to accept the new direction and try again. There is no sadness or regret, and no grief over the loss of control.

Initially I resisted this teaching. It was hard to surrender my fate to what seemed like pure chance. But then I found a wonderful story. To practice surrendering control, the early Muslims left their horses untied. But not until they learned to "tie the horse and then surrender" did they truly give up control. That's when I learned what I came to call *committed acceptance.*

Remember!

Take the responsible action first, then release the need to control.

The beauty of committed acceptance is that it doesn't take away from your chances of success. Quite the opposite: it's not your expectation of success that drives results; it's your diligent action that delivers them.

Here's a little riddle that conveys the same lesson. My drive from home to work has no traffic lights. When I drive at the speed limit it takes exactly eleven minutes. On Monday, I expected to get to work in nine minutes; on Tuesday, I expected the drive to take fifteen; on Wednesday, I was in full control and on time for my first meeting; on Thursday, I was stressed, worried, and late; on Friday, I really enjoyed the drive. Each day I acted as I should and drove exactly at the speed limit. How long did it take me to get to work each day of last week?

Eleven minutes!

If you take exactly the same steps, you will always reach exactly the same outcome regardless of your expectations, frustrations, pressures, or joy. The quality of your actions should not vary, and neither should your persistence in the face of challenges.

I made practicing committed acceptance my priority. I focused on doing the best I could every minute in every situation. I kept aiming high but remained emotionally detached from the results. If I missed a target, I looked back, learned, and tried again as if nothing was lost—because nothing really had been lost. At work I realized that I couldn't control every one of my employees, especially the really, really smart ones. I couldn't force a customer to buy my product, and I couldn't get the engineers to build it to my specs, or finance to price it as I wished, or legal to offer easy terms. Everyone had a different objective, and I needed to bring them all along. I learned to do the best I could without exerting, or expecting, full control.

In my personal life I make it even simpler: I plan, but I don't attempt control beyond the span of the present moment. Like Ali, I've learned to do the best I can in every situation and trust that all will work out fine.

Your Attitude

While actions are the visible levers of achievement, attitude is the true game changer. Consider the story of Tim and Tom.

When it was time to wake up, Tim hit the snooze button twice, then realized he'd be late for his nine o'clock appointment. He jumped out of bed in a panic, only to realize that it was raining so heavily that he'd certainly be even more delayed. He skipped his coffee and jumped in the car, looking shabby and feeling grumpy. *This is going to be a lousy day*, he thought. Already tense, he let his stress get the best of him, and he started switching lanes, banging on the steering wheel, and shouting "Come on!" Then—BAM—the car behind him rear-ended his. It was nothing more than a fender bender, but he jumped out of his seat, charged toward the other car, and violently banged on the hood, screaming and swearing in anger. Tim's behavior was so out of control that he ended up spending the night in a jail cell. *I knew this was going to be a lousy day*, he thought. And ignoring the impact of his own attitude, he went on to think, *All because it rained*.

Now let's replay the same sequence of events—snooze button twice and rain—only this time it's Tom realizing that he won't make his nine o'clock. So he brewed a good cup of coffee, showered and shaved and dressed in his favorite shirt, then grabbed a CD of Tina Turner's "I Can't Stand the Rain" because he knew it was going to be a long, slow commute. *I love the rain. I'm going to enjoy today*, he thought. He called his appointment to apologize and found out that she too was stuck in traffic. He sipped his coffee while he inched along, tapping his fingers rhythmically to the music, feeling really great. Then—BAM—the car behind him tapped his rear bumper. Curious, he got out and realized it wasn't a big deal. He smiled at the other driver and said, "Are you okay?" Relieved, she got out of her car, and she was stunning. "Hi, it's good to meet you!" he blabbered. She laughed and said, "Good? But I just crashed your car!" "Oh, but it's a good crash," he replied. Then she laughed again and said, "I love the song you're playing." And so it went. It felt like a moment from a romantic comedy. The rain added to the romance, and before long they both knew it was going to be a memorable day—all because it rained.

What's the rain got to do with anything?

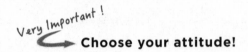

Choose your attitude!

I once attended a training course on change management in which we spent most of the time watching the movie *Apollo 13,* the one in which Tom Hanks plays astronaut Jim Lovell, whose mission was scheduled to land on the moon until an oxygen tank exploded two days after launch. Suddenly success was no longer a question of a successful moon landing but whether the crew would ever make it back to Earth.

There's a long moment of silence as the tension mounts, and then the silence is broken by Lovell's calm, confident, *almost cheerful* voice saying, "Houston, we have a problem." There is no trace of panic. If you'd just walked into the room, you'd think his problem could be nothing more than a flat tire. He then proceeds to describe what happened and asks for advice on how to handle the situation. Step by step, the crew devises an ingenious solution, and, eventually, they make it back home.

This concluded the training. The instructor had nothing more to say because Lovell's calm and assured attitude was what we were there to learn.

Life is bound to deal you a few bad hands now and then. You don't need to make a big deal out of every unexpected turn of events. Your path may be rerouted, but nothing is lost unless you decide to quit. Through it all, arm yourself with the right attitude. As Oscar Wilde said:

Very Important !

"It is all going to be fine in the end. If it is not yet fine, then it is not yet the end."

Thriving Out of Control

There is nothing wrong with planning and trying to assume control. The way we react when something unexpected happens is where we go off track. When things change, we react by trying to exert more control in an attempt to get back on track. What we should do instead is look at the current situation with an open, fresh perspective and attempt to use the new events in our favor, despite their having taken place beyond our control.

In algebra, when a parameter is irrelevant to the solution of an equation, we cancel it out. For example, If $A+C=2B+C$, it doesn't really matter when solving the equation what the value of C is. A will always be equal to 2B regardless, so we treat C as if it doesn't exist and solve the rest of the equation. C represents the parameters you can't control.

In the movie *Life Is Beautiful*, Roberto Benigni plays a Jewish father arrested along with his son during World War II and sent to a concentration camp. Despite the misery, sickness, and death that surround them, the father convinces his son that the camp is actually a complicated game in which performing certain tasks will earn them points, and whoever gets to one thousand points first will win a tank. Viewed within the context that it's all a game, the guards are mean only because they want the tank for themselves, and the dwindling numbers of children (who are actually being killed in gas chambers) are only hiding in order to score more points. The father realizes that the suffering his son is being exposed to is inevitable; the best thing he can do is be happy and playful to help his child survive.

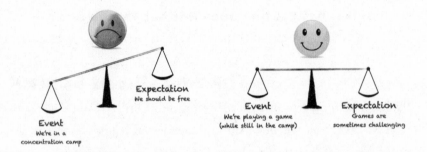

From time to time we may all face a hardship that is inescapable. If there's nothing you can do to change your current circumstance, then cancel the surrounding environment out of your Happiness Equation and solve the equation by using the rest of your life.

When life gets tough, some of us feel that we've lost the game and life has won. But life isn't trying to defeat you. Life isn't even a participant—the game is yours.

We're each handed a set of cards—some good, and some not so good. Keep focused on the bad ones, and you'll be stuck blaming the game. Use the good ones, and things become better: your hand changes and you move forward.

My happiness idol, His Holiness, the Dalai

Lama, is a shining example of this kind of commitment. He has been exiled from his country. His people have been subjected to violence and have had to endure years of lack. Yet with wisdom and peace he has done what is within his control while accepting what is not. In so doing he's become an ambassador for happiness to the whole world.

My Attitude

In my case, nothing has helped me through the tragedy of losing Ali more than the understanding of the Illusion of Control. Is there anything I can do to bring him back? Was there anything I could have done to save him? Is there a way to spend a minute more together? Would any amount of grieving be rewarded with a chance to see him again? No ☹!

I set the expectation side of my Happiness Equation based on the truth: Ali left. All I can control now is my actions and attitude. I choose to be positive and grateful for the years he blessed us with his presence. I choose to honor his life with my actions. That's within my control. I will turn sadness into happiness and do what I can to continue his life through the life of those who will reap the benefits of the contributions I make in his honor. I will give away the investments I planned for him and the fancy cars he never enjoyed. I will turn tragedy into smiles. Whenever I feel down or defeated, I hear him singing a line from the theme song from one of our favorite video games: "There's no sense crying over every mistake. You just keep on trying till you run out of cake." This is how the game of life is won. This is all I can control.

Chapter Eight

Might as Well Jump

6 Grand Illusions

You Are Here

o Thought
o Self
o Knowledge
o Time
o Control
o Fear

7 Blind Spots

o Filters
o Assumptions
o Predictions
o Memories
o Labels
o Emotions
o Exaggeration

5 Ultimate Truths

o Now
o Change
o Love
o Death
o Design

Happiness

Joy

I've never met anyone who's not afraid of something. Have you? Some may hide it well, keeping a brave face, and some may not even know the motivation for much of their actions is fear. But everyone has at least one fear that rules their life and limits their freedom. That's because fear is the granddaddy of all illusions, the one that rules them all.

Even if you're the president of the United States, the most powerful person in the world, there is something that you're afraid of. (By the way, it'd be cool if you were reading my book, POTUS ☺.)

I believe I can help you conquer your fears, but you need to be open and honest with yourself. The path is

complicated, but I'll take you through the process step by step so you can finally live free from the anxieties holding you back.

Admit That You're Afraid

Many people don't realize the true extent of their fears, how deep and how wide they are. No problem can ever be solved until it is precisely identified, so the first step to dealing with your fear is to admit that you're afraid.

Imagine taking everything you were supposed to do next week and deciding not to do it. Can you stop going to work? Why not? Is it the fear of losing your source of income? Or are you worried about what they will think of you? Can you keep the doors unlocked? Why not? Are you afraid someone will walk in and steal your TV? Are you scared of the threat to your own life? Can you stop talking to that one annoying friend? Can you stop taking your supplements? Can you drop your health insurance? Can you take your kids out of school? Can you give away all your money? Why not? Fear!

It's normal to be afraid. What's wrong is to behave as if you're not, because that leads to wrong decisions. We tend to come up with reasons to explain why the choices we make aren't driven by fear. If your relationship isn't working but you're unable to end it, you'll find a perfectly justifiable reason to persist in your suffering. "I want to stay with my partner because of love," you say. Ask yourself what you would do if someone else came along, someone rich and famous and incredibly good-looking, mature, and kind, who loved you dearly and gave you all you needed. What if there was nothing to fear? Would you stay in your relationship? If not, it's not love; it's the fear of losing what you have and the fear of being alone.

Fear isn't always so obvious. It comes in many different forms. Anxiety is a direct derivative of the fears we let linger. It results from thoughts in our head or from projections of imaginary events. Frustration is driven by the fear that further attempts will not achieve your goal and that not achieving it will result in a worse future than simple failure. Disgust is the fear of interaction with something that represents a prospective displeasure or harm. Grief is partially driven by the fear of how life will be after the loss, fear for the safety of the loved one in light of the mystery of death, and fear of one's own death. Embarrassment is the fear of rejection due to one's past actions. Envy and jealousy are driven by the fear of being less than another. Pessimism is the fear that life is always out to get you, that future moments will be worse than the present. Every negative emotion you will ever feel will have traces of fear all over it.

Whatever it is, there's always something that frightens each of us—or at least *worries* us—enough to keep us locked inside a routine, losing out on experiencing all the different flavors of life.

 But we don't admit it. We think fear is a sign of weakness. It makes us feel vulnerable. We act strong, puff out our chests, and hide our fears. We practice our disguise so long that we believe it. Think about it, though: when is a puffer fish fully puffed? Being puffed isn't a sign that it is brave but a sign that it is afraid, very afraid.

When you find it difficult to admit your fears, ask yourself a different question: *Are you free?*

This question helped me uncover my fears one by one. And there were many. I'm no longer ashamed to admit it—it's part of being human. Over the years I've managed to overcome some fears, but I still struggle with many others. Chief among them is a deep fear of failure. It drives me to blow things out of proportion and set unrealistic targets for myself. In my personal relations, I'd go out of my way to make sure that my loved ones are cared for and visibly happy all the time. If they're not, my fear takes over and I consider it a sign of my failure. I convinced myself for years that I'm just a perfectionist, but that's a lie. I'm afraid of failure.

There, I said it. I admit that I'm afraid. Now it's your turn.

It's not rocket science: if there's something you want to do but aren't able to, then you're not *free* even though you're not in a physical prison. Think about the invisible walls of your captivity. Call them anything you want—or just call them fear.

Understand What Fear Is

Every fear originates from a conditioned response. Most of the time our conditioning delivers a subtle but sufficient dose of fear that keeps us from being totally free, even if the original reason for the fear no longer exists, and even as the underlying reality of the threat becomes insignificant.

In the first half of the twentieth century, psychology was dominated by the study of conditioned responses. In 1924 the father of the behaviorist school of psychology, John B. Watson, famously said, "Give me a dozen healthy infants and my own specified world to bring them up in and I'll guarantee to take any one at random and train him to become any type of specialist I might select—doctor, lawyer, artist, and, yes, even beggar-man and thief." [1]

For Watson, this was more than talk. In 1920 he ran an ethically dubious experiment to demonstrate "classical conditioning" on a nine-month-old infant. Little Albert was shown a white rat, a rabbit, a monkey, and various masks. With no conditioned fear yet in him, Albert interacted positively with each of them. His favorite was the white rat, until Watson hit a steel bar with a hammer behind Albert's head when the rat was presented. The sudden loud noise would cause Little Albert to burst into tears. The sequence was repeated seven times over seven weeks; by the end Albert only had to see the rat to immediately show every sign of fear. He would cry and attempt to crawl away even in the absence of the loud noise. A lifelong fear was created.

I have personally seen a phobia develop in my charming daughter. Aya must have been around a year old, sitting on the floor playing peacefully on a summer night. We had left the windows open, and a flying cockroach came in and landed right in front of her. With no previous conditioning to be afraid of cockroaches, Aya grabbed it as if it were just another toy. She looked at Nibal and waved her hand, totally happy with her "new toy."

To Nibal, though, cockroaches are more dangerous than a nuclear blast. And her reaction was more alarming than that hammer hitting a steel bar behind little Albert. She screamed in horror, started crying, and shouted my name for help. Seconds later, the unwelcome visitor was gone. There were no human casualties, but Aya was conditioned. Years later, when I attempted to play a practical joke on Aya that related to cockroaches, her fear had grown stronger. She screamed, cried, and ran away. She remembers this as one of my worst behaviors still today. Sorry, my wonderful Aya.

Name Your Fear

Acrophobes are afraid of heights; claustrophobes are afraid of enclosed spaces; nyctophobes fear the dark; and trypanophobes fear injections or medical needles (one of my terrors). Because those fears relate to tangible somethings they're highly visible and easy to spot. But how about the fear of social rejection?

With some fears there's a much more fluid definition of what we dread, and that makes them much harder to pinpoint. There are so many hidden fears. We live with them as they eat us up from the inside. Some people fear not having the resources to buy what they need; they get hooked into an endless attempt to accumulate as much as they can, but they never feel secure, regardless of how much wealth they accumulate. Others dread losing their freedom; this may include losing the freedom of physical mobility, losing the freedom to express one's opinion freely, or losing one's ability to freely make decisions because of external controls such as a boss, a corporate structure, or even a committed relationship, such as marriage.

Some fear the unknown, failure, or not meeting expectations. Some fear loss of control; others fear loneliness, social rejection, or ridicule. All of us fear death, and consequently many of us fear aging. And the list goes on and on.

What are your fears? If you find it hard to admit them, that may be due to another overarching fear: *the fear of facing your fears.*

At a fundamental level, many of us fear finding out who we really are and what about ourselves we need to fix. Denial allows us to procrastinate as we learn to limit our lives to cope with our fears. Is this one of your fears? If it is, then it's time to face it. Time to admit you're human. And like all humans, there will always be some fear to face.

Your Brain's Fear Games

The fear of facing your fears is just one of many games your brain plays to ensure that you're fully obedient and under control. As the games commence, your brain attempts to build a logical construct that hides the real source of your fear, which you'll find originates from another, well-hidden, deeply buried pain. Our fears are tricky to uncover because they hide and morph.

In its purist form, fear is a defense mechanism that is triggered to warn you of proximity to harm. Fear alerts you so that you can take the actions necessary to avoid the potential of suffering pain, physical or psychological. But pain itself is just a mechanism also controlled by your brain. The pain of touching a hot stove doesn't happen in your hand at all. Instead, a signal is transmitted to your brain, which labels it as pain. Accordingly, scientists can simulate the experience of pain merely by stimulating certain parts of the brain. This makes pain just another form of thought.

In that sense, you could consider that pain isn't real because an identical event—touching a hot stove—may produce a drastically different response. Individual tolerance of pain varies depending on the situation. As we grow beyond early childhood, for instance, we can withstand hunger a lot longer than we could as infants. Clinical studies published in the *Journal of Psychosomatic Medicine* asked participants to submerge their hands in ice water and found that the promise of financial compensation could induce them to suppress pain and keep their hands submerged longer than those without the prospect of a reward.[2]

Because pain is just a thought, your brain can ignore it, and you can learn to suppress it. That's what long-distance runners do. Sometimes you can even learn to enjoy pain. Muscle soreness after a good day of exercise is something we learn to like because it's a feeling we associate with growth and improvement.

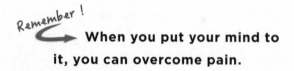

Remember !

When you put your mind to it, you can overcome pain.

Everything that applies to physical pain also applies to emotional pain. We tolerate emotional pain differently, depending on circumstances, but most of us can learn to suppress it or even use it to our benefit. The pain of rejection, for instance, is much worse for a teenager than for someone older and less insecure.

So why is it that we don't usually suppress emotional pain? Because, just as is the case with physical pain, our brain uses emotional pain to keep us away from harm. The difference is that physical pain can't be generated by our brain on demand, but it can regenerate emotional pain by using incessant thoughts. And that leads to suffering.

Our brains replay every painful memory from the past and every possible scary scenario from the future over and over, just like a complex computer simulation, in an attempt to scare us away from threats before they can happen and regardless of the probability of their happening at all. Every time our brains find possible threats in our simulations, we associate them with a form of fear, and while the threat themselves might not be very significant, our brains exaggerate the fear.

Let's say you have a fear of public speaking. If asked why you have that fear, your initial response might just be "Because." But if you dig deeper, past the brain's defense mechanism, you'll discover where that fear really stems from.

What is it that you are really afraid of?
I am afraid of saying something silly in front of a large audience.
And why would that scare you?
I am afraid of being judged or ridiculed.
And why would that scare you?
Because I may get rejected as a result.

Keep going until there is nothing more to discover. You will have uncovered the layers of unnecessary fears we suffer due to a brain mechanism that I learned to call the *safe model*.

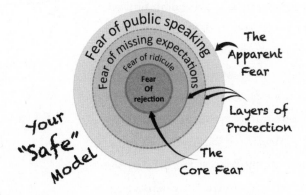

To avoid a specific fear, your brain tends to look for every possible threat that may trigger it—every painful experience from the past and every possible scenario from worrying about the future. It registers the threats it finds as more things to fear. It's just *safer* this way, your brain thinks, but is it?

Every new fear causes more insecurity more often. Instead of one fear to deal with, you now have many. The compounded effect is significantly intensified. With more to fear, your brain tries harder to keep you safe. As a result, the vicious cycle continues: *more* fear calls for *more* layers of protection.

In a hopeless attempt to keep you as far from harm as possible, your brain builds what it believes to be a *safe model*, an elaborate structure with a large number of scary scenarios to worry about and more barriers—fears—to defend you against them. We try to plug every hole and patch every crack. But what we build is a rickety structure. *The more we build it up, the more threatened we feel and the more weak spots we expose.* It's a matter of simple mathematics: the higher the number of vulnerable points, the more frequently one gets compromised.

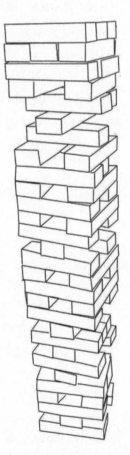

Layer over layer, our defense structure becomes the main source of our fragility. The mounting pain becomes disproportionate to the underlying reason for our original fear. It all becomes unbearable, fear becomes a way of life. As we try harder to stabilize and expand our safe model, we fail because some random event is bound to threaten one part or another. Every time this happens, it acts as confirmation that we had a good reason to be afraid, and so the vicious circle continues. Life truly becomes a long horror movie, one that comes with no advertising breaks.

Very Important !

→ There is no safe model.
The harder you try, the more you will fail.

Once we build our model, we have a hard time getting rid of it. We make that model the foundation of our expectation in the Happiness Equation and compare life to it as it unfolds. The two never match. We are disappointed, we suffer, and we become anxious that nothing is ever safe.

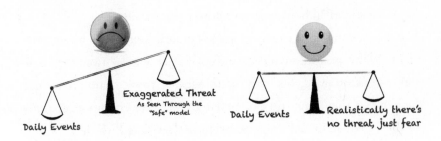

Simple things can easily become major threats because eventually, through all of those protective barriers, they lead to your biggest fears. "If I speak in front of these people, I'll stumble over my words. If I stumble over my words, they won't take me seriously." "I don't like the heat. It'll spoil my makeup. That will make them judge me. If they judge me, they'll reject me." Something as benign as a warm day becomes part of the fear of rejection. *Everything* becomes an intrusion to your safe model. We become perpetually unhappy, not because life is unfair but because our expectations are totally blurred by the Illusion of Fear.

Yoda, the wise Jedi master of *Star Wars*, sums it all up in one statement: "Fear is the path to the dark side. Fear leads to anger. Anger leads to hate, and hate leads to suffering." I love Yoda.

The only way to escape the vicious circle is to collapse it at the core with all its parts and all at once. Facing your fears one by one may seem hard, but it's easier than you think.

Take the Vow

Many of us become content with suffering and come to believe that this is how life is supposed to be. We put up with suffering in fear, mostly not knowing exactly what it is we're afraid to face. The first step on the path to live free is to look your fears in the eye and acknowledge them. Instead of hiding away, you need to face them.

Do you know how elephants are kept in captivity? By a flimsy little chain. Those nine-thousand-pound giants could snap away the chain with no effort at all, but they don't because those chains restrained them when they were infants, and they became conditioned. Early on they tried and tried to break free, but they failed, so they stopped trying. We behave the same way too. We exaggerate our fears and stop trying to break free.

By now I'm almost certain that your brain is telling you, "But fear *can* be good. Don't believe this Mo guy. Our fears are what keep us out of harm. There are positive aspects to fear."

No! There are none. What keeps us alive and propels us forward are our *actions*, not our *fears*. Fear, if anything, paralyzes us. It blurs our judgment and blocks us from making the best possible decisions.

Fear of failure doesn't drive our best performance. All it does is add anxiety. What truly drives us to success is our hard work. And you don't *need* to be afraid to work hard. Looking back I realize that when

I achieved any kind of success, my fear of failure often took over and caused me to be fearful of the next stretch of the journey, and I never gave myself the chance to enjoy even the best moments of my life. My fear took away my happiness throughout the journey and even when it was time to celebrate.

Remember !
There are no positive aspects to fear. It's your actions and not your fears that keep you safe.

The thoughts that lead to fear are always anchored in the future. Your brain tries to make you believe that the next minute will very likely be worse than the present unless you do something to protect yourself. When afraid, you believe that life is trying to trick you and that you're in danger unless you do something to stay safe.

But do you truly believe that life, with all its might, with its endless resources and infinite connections, is designing its next move just to get you? Do you truly believe that the orbiting of the planet and the cycles of the unfolding lives of more than seven billion people are here just to scare you? If life were really out to get you, do you believe that your flimsy protection would keep you safe? Well, you're kidding yourself: you'd be toast.

Remember !
The only thing life *wants* is to be experienced.

Life wants you to sample every flavor it can offer you. Sour isn't worse than sweet; they're just different. Life keeps trying to capture your attention while you try as hard as you can to shut it out. It keeps presenting you with experiences, some to enjoy and some to learn from as you develop and grow, but you stay locked up inside your fears, refusing to live them.

Be honest now: How often did your worst fear come true, and how often did it not? How often did a fortunate twist of fate give you more than you had hoped for?

The future will be better than you expect it to be. It always has been. **You would not have been here if your present matched your past fears, would you?**

As we get locked in cycles of distress about the future, we forget that fear itself is proof that we're okay. Think about it: **If you can afford the brain cycles to worry about the future, then by definition, you have nothing to worry about right now.**

Remember! → **Right now, you're okay.**

Many kids cry on the first day of kindergarten. They stomp their feet and scream because they're afraid. Then, a few hours or days later, they're okay. They even like it. What's the magical transformation that takes place? Does kindergarten change to match their expectations? Not at all. Nothing changes. But when they face their fear they realize that playing with other kids all day is not bad after all.

We do this over and over. Some of us fear confronting a bully; some fear making our first presentation; some fear leaving a bad relationship; some fear walking up to a stranger and saying hello. But when we act in spite of our fear, we realize that there is nothing to fear. It's challenging at first, but once you overcome your fear, you realize it was well worth the effort.

Are you ready to take on that challenge?

Take the Leap

To conquer your fear you need to put yourself face to face with it.

The easiest way to short-circuit all of your brain's fear games is this: Once you know what your fear is, **commit yourself to facing it**. If you fear public speaking, find the next opportunity and volunteer to be a speaker. Put yourself beyond the point of no return. *Don't think. Just do it*. It will be fine. I promise.

Here's a series of simple questions that will guide you through the quest to overcome your fears. Because they help you see what your brain is hiding from you, I call this list **The Interrogation**.

What's the Worst That Can Happen?

The minute you ask this question, your brain will jump into hyperdrive and start imagining a thousand and one horror stories. Don't resist. Play along. Let your brain run with it. Then boil it down to the one truly worst thing. We aren't interested in the other thousand scenarios, only in the absolute worst, but realistic, outcome. Use this fear as an example: *What is the worst that can happen if I speak publicly?*

I might be extremely boring and put everyone to sleep.

That's not too bad. What's worse than that?

They might laugh.

Is that the worst?

No, they might boo me off the stage.

Okay, that would be bad. Can it be worse?

Of course. My boss could be in the audience and I could lose my job.

Aha, now we're talking. Can you think of any worse scenarios?

Yeah, a mad sniper in the audience may decide to shoot me.

Let's be realistic. Seriously, what's the worst that can happen?

I told you, they'll boo me off the stage and my boss will be there
and I'll lose my job.

Okay. I think we're there.

This question makes you visualize the worst-case scenario for your fear. I know you must be hurting now just thinking about it; please forgive me. But I have great news for you. Identifying the worst-case scenario helps you reach the bottom.

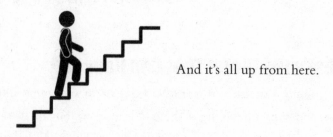

And it's all up from here.

Let's start climbing up. The next question may surprise you.

So What?

This question is the turning point away from fear and into the direction of courage. So what if I lose my job? Will my life end there? Will I starve to death? So what if they boo me off the stage? Will I cease to exist? Besides the thought in my head called shame, is there really any damage that happens as a result of being booed? If that's the worst-case scenario, can you see that if you ignore the pain associated with it, you can survive? Let's continue to climb.

How Likely Is It?

Honestly, how likely is it that the worst-case scenario will happen? Has it ever happened to you? How often have you seen it happen to someone else? How many times did you see horrible speakers on stage, and how often did you see them being booed? Do the math. Keep climbing.

Is There Anything I Can Do Now to Prevent This Scenario?

That, my friend, is my favorite question of all. This is when you turn your fear into action.

Get to work and prepare like crazy. Give your speech in front of the mirror, your spouse, and the dog a hundred times. Do it until you're totally comfortable. Then do it again. Being prepared will, at the very least, further reduce the likelihood of the worst-case scenario and help you feel calm that you gave it your best shot.

This, by the way, is the time when your brain might start to fight back, asking, "Why are you doing this to me? My life was so easy before you started reading this book."

Ignore it. You're almost there. Onward.

Can I Recover?

It gets even more interesting when you ask yourself: What if the tiny probability of my scariest scenario comes true and I do get booed and I do lose my job? Is that a situation I can recover from? Could you perhaps reduce your expenses for the next few months? Will you eventually find another job? You will, and one with a better boss, I hope.

It'll be a bit unpleasant, I admit, but that will pass just like every other unpleasant experience in your life so far has passed.

Feel any better? We've followed the right thought process and in doing so unmasked the fear that your brain locks you into. Underneath that scary mask, there's just a little harmless kitten. The rest is our imagination. The scariest scenario isn't going to be the end of your life. When you take action, you reduce its possibility even further. And if it ever does happen, you will always find a path to recovery. What a relief!

But wait, this gets even better. There are more stairs!

Your brain's inclination is to think about what might go wrong. That way it can plan ahead for threats and ensure your survival. Two more questions can help you shift your thoughts away from all the bad things that scare you to all the good things that await you, so you can take the big leap from your fear.

What Will Happen if I Do Nothing?

Now *that* is a good question. What's the price of the status quo? Is that a price you are willing to pay? What's the price you would pay if you stayed in that abusive relationship? What's a better path through life? Living alone, even for the rest of your life (worst-case scenario), or being abused? What would happen if you remained stuck with a job that's bringing you down? What's the price you have to pay if you don't face that bully or acknowledge that you need to rework your finances?

I can promise you this: what you suffer when you stay in fear is almost always more damaging than facing your fear. This is the reason your brain exaggerates your fears so much: to make them a bigger threat than the pain you currently feel. It has to—otherwise you'd get rid of them easily!

Next is the best question of all.

What Is the *Best*-Case Scenario?

You know those scenes in action movies when the secret agent flips a switch and the entire lair of the villain crumbles and explodes in a spectacular display of fireworks? This question does exactly that to the fortress of your fear while your brain is looking on with dismay, puzzled by how you managed to isolate it into this corner, and suddenly afraid that it might be running out of tricks. Hit it while it's on the ropes. Win this match.

What's the best that can happen? *That* is the question to ask.

What if you left your job and went back to playing the piano? What if facing the bully made him go away? What if all the stars aligned? Would you write the next Harry Potter book? More important, would you be happy? Things often do work out. Why should you let yourself miss out on the upside?

The cost of doing nothing is often higher than the cost of facing your fear. And when things work out, the upside is absolutely worth the risk.

Realizing that your fear is exaggerated makes it easier for you to decide to face it. Picturing what awaits you on the other side gives you the energy you need to get up and do it. It gives you the resilience to withstand the pain associated with the process in hope of a better future.

Enough of your brain's doomsday scenarios! Refuse to take them anymore. Live every minute of your life totally optimistic about the next moment. Face your fears, one by one, and wipe them out of existence. They were never real anyway.

It's Time

I have found that when you evade your fears, they pop up to face you. Like a wise teacher, life will test you, fear by fear, to see if you're ready to move on to the next lesson. Once you conquer a fear, the test goes away and you never have to face it again. But if you hide, the test—the fear—keeps popping up to haunt you along your path.

Like everyone you know, I refused to admit my fear to anyone, including myself. I pretended I was brave—and I was lying. I feared failure. So I kept pushing myself. Succeeding as a businessman was an answer to my fear. Close a bigger deal, and you're successful; fail to close the deal, and you're a failure. I spent most of my time working and was paranoid about making any single mistake.

I kept my fear alive, and so life—the ultimate teacher—took charge and put me through the test. I had to face my fear when once I disagreed *royally* with one of my managers. The situation came to be beyond repair and I was very close to leaving—or being asked to leave. The pain was all too real. To be out of a job is the ultimate form of the failure that I feared. And that's when I realized that a change would be good. I chose to walk right into the center of my fear. I found joy in the freedom that my willingness to leave gave me. I knew then that if I lost my job, life would still find a way. So I left, and that's just what life did. Once my fear vanished, the test vanished along with it. Fast-forward and today I love the work I do. There was nothing to fear.

I wanted the best for my family and had no bigger fear than missing their expectations. I loved the comfort that money gave them, and so I grew a fear of losing it. I learned to save it and invest it. I nearly worshipped it until, one day, I made a major investment mistake and came close to being wiped out. Life put me face to face with my fear, and I realized it wasn't that scary. I realized that I needed much less money than I'd imagined, that my family's expectations of me were much less than I'd aimed for, and that if it all went away, life would still find a way. I felt liberated. Once I was no longer fearful, the test went away and I never had to worry about money again.

Test after test, my fears faded away, until, for a while, I felt that I lived fearlessly. I had a lot to lose but nothing that I feared losing. There was nothing that I cared about, that anyone could take away from me. It felt amazing.

And then Ali left.

There was never a bigger fear. There was no one and nothing in life I protected more. I kept it hidden deep inside, but losing one of my children has always been my true nightmare.

One last time life threw me into the center of the arena to face my biggest horror. The pain was unbearable. It still is, but in the process life wiped away my last fear. There is nothing more that can be taken away. With that one last move on the chessboard, I win, or perhaps I lose. Either way, there will never be another fear.

Even as I pray for the well-being of Aya, the sunshine of my life, I hope that this test is done. There's no need to take the courage test because I already passed.

Death is the biggest fear of them all, and learning to face your own death is the ultimate form of facing your fears. When Ali left, I died, and I say that in the most positive sense. Life finally fit into perspective. I have an overwhelming feeling of peace. There is nothing more to lose; there is nothing more to fear. Eckhart Tolle says this is "to die before you die," to live life knowing that because one day it'll all be gone, there's really nothing that you have, and so nothing you have to lose.

Like a marathon runner, I reached my pain threshold when Ali left. I now know that the next step is just another step on the path, until I reach the finish line in peace.

I cry every time I remember that the price for my freedom was his life. But Ali too has found his path. He too is at peace.

I know that you are happy wherever you are now, Ali. It'll just be some more glorious days to go before I get that hug that I miss and get to hear you say our usual greeting, *"Ezayak ya aboya."* Until then I'll try to live fearlessly. Only then will the journey be complete.

There isn't a single day in life worth living in fear. Life will bring you face to face with your fears unless you decide to pass the test before it's brought to you.

Very Important !

**→ Learn to die before you die.
It is time to face your fears.**

Part Three

Blind Spots

7 blind spots affect the way our brain processes information and blur our perception of reality. To ensure our survival, the seven blind spots are combined with the brain's tendency to be pessimistic. This interferes with our ability to solve the Happiness Equation, thus making us suffer needlessly.

Part Three

blind spots

Chapter Nine

Is It True?

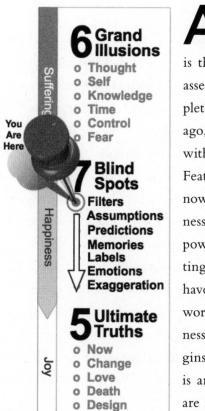

At the root of our challenging relationship with our own brains is the fact that it's a device that was assembled, tested, and (mostly) completed hundreds of thousands of years ago, in a vastly different environment with vastly different requirements. Features that were once advantageous now drag on our capability for happiness. Despite its immense processing power, the human brain is still spitting out solutions for equations that have little to do with our modern world—and less to do with happiness. Because of its evolutionary origins, the world your brain deals with is ancient, murky, and terrifying. So are its strategies. If we are to use this device properly, we need to adapt its programming to match its new operating environment. But first let's see how it all began.

On the Origin of Blind Spots

A branch on the bush a few steps behind the Cro-Magnon hunter shakes ever so slightly. The sound catches the skilled hunter's attention. He waves his hand to his gang, instructing them to lay low and stay quiet while he investigates the source of the sound. He squints his eyes, sharpens his hearing, and *filters out* every other sensory stimulation. That bush gets his undivided attention. Everything else fades into the background.

The wind is blowing from behind him in the direction of the bush. He reckons that this is why he couldn't smell the beast he fears. This is the game plan beasts follow as they attack. Clearly this is a clever predator, a tiger perhaps, and from the height of the branch that moved he further *assumes* it must be a large one indeed.

In the dead quiet, the hunters hold their breath. The bush stops moving, an indication that the beast knows it's been noticed. In his mind, the Stone Age hunter *predicts* an imminent battle. He imagines with precision the angle and speed of attack. The attack is just seconds away, he's certain, so he waves to his pals to take a few steps back.

His caution is based on painful past experiences. Since he first ventured into the jungle to hunt alongside his father, many fine hunters became prey to a wild beast in careless moments. Though many moons have passed, he *recalls memories* of how the beasts pounced, threw their victims to the ground, and tore the muscles from their bones. He lives that memory as if it were happening in front of his eyes, and his heart starts to race.

There isn't a moment to waste. Trying to process the fine details to further analyze the situation would eliminate his chances to escape. The risk is too high. He needs to make a snap decision, so he *labels* the situation a clear and present danger. When his life depends on it, speed matters much more than accurate investigations.

He *feels* an overwhelming *emotion* of panic. His brain imposes this state by flooding his body with adrenaline to prepare him for a fight-or-flight reaction.

As the panic takes over, his brain *exaggerates*, seeing every possible scenario as far more dangerous than it actually is. This could be a pack of tigers, he thinks. We could be surrounded. There's no point attempting to fight; we're all going to die. More branches move violently. In a split second, he instinctively turns his back to the bush and prepares to run—just as a few birds fly off. A little sheepishly the hunter looks up to the sky as he realizes his tiger is nothing more than a flock of birds. Who cares if the past few minutes were a bit stressful, his brain thinks. At least we're still alive.

For millennia, our brains have been equipped with the seven incredible features I've just highlighted: filters, assumptions, predictions, memories, labels, emotions, and exaggeration. Yes, these tendencies may have ensured the survival of our species long ago. And our ancestors didn't begrudge the discomfort those features caused them because they navigated an extremely hostile environment. For them, it made sense to assume the worst because the worst frequently happened.

As we developed civilization and drove the tigers away from our cities, swapped our hunting grounds for the jungles of the workplace, clubs, and malls, we have continued to rely on these seven features. Yet we seldom stop to ask how effective they've become in this "alien" environment. Just as a screwdriver can be used to tighten a screw or to poke us in the eye, those survival features can be turned into blind spots that work against us and make us unhappy, especially when they combine with another ancient tendency that's the core characteristic of the brain.

A Tendency to Be Grumpy

The way our brain operates reminds me of my first car, which was an old, battered, used car that was all I could afford. Often that car would suffer from one of many mechanical issues—the spark plugs malfunctioned, the starter coil failed, and the radiator leaked. On top of that the car never drove straight because the wheels needed alignment. That car was a wreck. At any point in time, one or more of the mechanical errors occurred, causing me problems. When the radiator leaked, the car overheated, and when the spark plugs failed, the engine jiggered. But even as I fixed some of those mechanical failures, the alignment issue kept shifting it to the left as I drove.

The same thing happens with our brain. Often one or more of the seven blind spots distorts our perception. Each would affect us differently as our brain tries to make sense of life. On top of the blind spots, however, an overarching tendency persists: the tendency to be grumpy and push most of our thoughts way out of balance.

After a while, it became dangerous to drive that old car unless I fixed the wheel alignment issue. Only when that was taken care of could I attend to the mechanical defects one by one. As a good mechanic would do, mine performed a thorough inspection of the car to assess the extent of the problem. Let's do the same with your brain.

The Inspection of Your Brain

Let's run your brain up on the lift and do two quick tests, Check and Track.

Check

Check out this image and make a note of what you see, just at a glance.

Did you notice the teddy bear in the little girl's hand, the book falling out of her bag, or the empty parking meter? What *did* you observe? Was it the Don't Walk sign, the boy running away, the speeding car, or the girl stepping into danger? Did you notice the imminent accident just about to happen? Most of us do.

Now zoom out and see the bigger picture. You will notice that nothing's going wrong. The car is actually parked, there's a policeman controlling the traffic, and everyone's safe. Why isn't that the scenario you anticipated?

Try to play Check as part of your daily life. In any situation you'll notice that your brain has a tendency to spot what's wrong and what can represent a threat. Only a lot less frequently will your brain pick up on what's going right or what's mundane. It's a bit like our hunter friend translating the movement of a branch on a bush as tigers rather than birds.

Track

Fold a piece of paper in half and mark one side with a plus sign (+) and the other side with a minus sign (−). Now observe the dialogue happening inside your head; notice every thought as it pops up throughout your day and add a tick mark on either side of the page depending on the kind of thought it is. Examples of thoughts that get a tick mark on the positive side are *Life is good to me; She'll love me forever; I'm beautiful.* Examples of thoughts that get a tick on the other side are *I don't like this job; Bad things always happen to me; He's such an idiot; I'm fat.*

Now count the marks. Is your brain producing primarily optimistic thoughts or pessimistic, judgmental, or critical (negative) thoughts?

Most people don't need to do this test for very long before they acknowledge that the majority of their thoughts are negative, cautious, judgmental, and pessimistic. Is that the case for you too? Don't be upset about that. We're all right here with you.

Ample research has shown that **we tend to think negative— self-critical, pessimistic, and fearful—thoughts more often than positive thoughts**. Psychologist Mihaly Csikszentmihalyi uses the term "psychic entropy" to indicate that worrying is the brain's default position.[1]

Raj Raghunathan and colleagues at the University of Texas conducted a study similar to the Track test. Students were asked to maintain a "brutally honest" record of their naturally occurring thoughts for a period of two weeks. The tally revealed that somewhere between 60 and 70 percent of the average student's thoughts were negative, a phenomenon known as the "negativity dominance."[2] Those ratios should not be taken lightly. Based on Deepak Chopra's blog post "Why Meditate?", this can add up to a whopping 35,000 negative thoughts per day.[3]

But our bias to negativity isn't limited to the sheer number of thoughts. **We also tend to give greater weight to negative thoughts when we make decisions.** The work of Roy F. Baumeister, Ellen Bratslavsky, Catrin Finkenauer, and Kathleen D. Vohs shows that people are more likely to make choices based on the need to avoid a negative experience rather than the desire to attain positive outcomes, a phenomenon known as "prospect theory."[4] This is why if a restaurant was once rated a single star and once five stars on Yelp, you will likely give more weight to the negative rating and decide not to go, although statistically the five star rating may be equally true.

We also dedicate more of our brain resources to negative information. Felicia Pratto and Oliver P. John of the University of California at Berkeley ran a study in which participants were asked to observe a series of words appearing sequentially on a computer screen. The words appeared in different colors, and each was the name of a positive or negative personality trait. The traits were irrelevant to the task, which was to name the color as quickly as participants could. Yet participants were noticeably slower to name the color when the trait shown was negative. This difference in response latencies indicates that participants devoted greater attention to processing the trait itself when it was negative than when it was positive.[5]

Another interesting finding was that participants exhibited better incidental memory for the negative traits than they did for the positive ones, regardless of the ratio of negative to positive in the set. This implies that **we tend to remember negative traits more easily**. As a result, **we tend to recall negatives more often**. When asked to recall any recent emotional event, we tend to report negative events more often than positive events. We also tend to underestimate how frequently we experience positives because we forget the positive emotional experiences more often than we do the negative ones.[6]

Socially, **we tend to offer more respect to those who are negative** than to positive folk. Clifford Nass of Stanford University argues that we view people with a negative perspective on the world as being smarter than those who are positive.[7] **We even have more negative words in our vocabulary** (the building block we use to construct our thoughts)—62 percent of all emotional words in the English dictionary being negative.

None of these negative biases is a coincidence. They're clearly reflected in the design of the brain. For instance, the amygdala uses approximately two-thirds of its neurons to detect negative experiences, and once the brain starts looking for bad news, it stores it into the long-term memory immediately, while positive experiences have to be held in our awareness for more than twelve seconds in order for the transfer from short-term to long-term memory to occur. Rick Hanson, senior fellow of the Greater Good Science Center at UC Berkeley says, "The brain is like Velcro for negative experiences but Teflon for positive ones."[8]

The evidence is overwhelming and I could go on, but the bottom line is this:

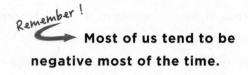

Remember!

Most of us tend to be negative most of the time.

So why is our brain so grumpy? To find out, we'll need to leave the workshop and get into the real world.

A Diligent Lawyer

Your brain often looks for what might pose a threat. Why would it do otherwise when its sole purpose in life is to protect you?

Imagine if the conversation in our ancient hunter's head went like this: "Chill, there are no tigers around here. Don't even bother checking. Walk into that cave, you'll be just fine." That kind of optimism would have allowed for a less stressful life, but probably one that was considerably shorter. When raw survival is at stake, it's better to be safe than sorry.

Your brain isn't there to encourage you; it's trying to protect you. That's why it often behaves like a diligent lawyer would. Given the task of protecting your business from every possible attack, good lawyers write hundreds of pages of contracts and legal documents that anticipate every little thing that could possibly go wrong. Most of it will never happen, but in the unlikely event that it does, they don't want to be called out as the one who missed it and put your entire business at risk.

Because they prioritize our survival over our happiness:

Remember !

➤ **Our brains tend to criticize, judge, and complain more often than not.**

They also tend to ignore the happy events, since those offer no survival benefit. This makes the majority of the conversation in most of our heads, well, grumpy!

This grumpiness represents a form of disagreement with life. It renders a view in which events contradict the expectations of a safe, unthreatened life. Factor such disagreements into your Happiness Equation and the result will be unhappiness.

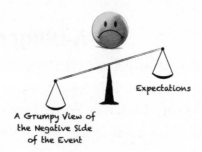

A Grumpy View of
the Negative Side
of the Event

With your brain's blind obsession to keep you alive, it conveniently ignores what's blatantly obvious: that **the negatives we face are the exception that interrupts a norm** of constantly flowing positives.

You don't believe me? Then answer this: What's the norm, health or illness? Good weather or typhoons? How often do you have to live through an earthquake compared to walking on solid ground?

Remember!

→ **Life is almost entirely made of positives.**

Ignoring the positive events makes for poor judgment. It's like the black ink on this white paper. Our eyes are trained to look for the black—the ink—but most of what we're looking at—the paper—is white. When you choose to focus on the white instead of  the black, you'll notice different angles and perspectives that are also there, perhaps in greater numbers. Stop the grumpiness.

Remember!

→ **Focus on the white of the page, not the black of the ink.**

The Whole Truth and Nothing but the Truth

Now let's get back to the seven blind spots. Remember how the hunter reacted to the movement of a branch? Compare that to the following common scenario.

As you walk into your office and put your things down, you inadvertently push a pencil off the desk and onto the floor. That event, on its own, is insignificant. Your brain, however, might launch into the following conversation:

I lost my pencil. I can't find it anywhere (*filter*).

I love that pencil (*emotion*). I can't live without that pencil (*prediction*).

It's my lucky pencil (*label*). We've had so many successful meetings together (*memory*).

Without that pencil I'll fail. My kids won't have anything to eat (*exaggeration*).

Someone stole it (*assumption*). It must be Emily (*assumption*). She's mean (*label*).

If I let this happen once, I'll be the doormat of the office (*prediction*). Today a pencil, tomorrow my job (*exaggeration*).

As you sit down to prepare your attack plan, Emily passes by and says, "Hey, you dropped a pencil."

It really is just a pencil !!

We've all experienced similar scenarios. Have you ever found yourself overreacting to a comment by a friend, only to discover that he didn't mean what you heard? Have you ever predicted a future disaster that had no basis of truth to support it?

In the scenario above, it's just a pencil, right? But the thought in your brain made it your doom. If our thoughts can turn such an insignificant event into such a serious drama, then perhaps we need to ask an obvious, though rarely asked, question:

Very Important !

→ How much of the constant stream of thoughts in my head is true?

There's no better place to answer this question than the one devoted entirely to finding the truth: a court of law. But this time we'll not let your brain stay in its comfort zone and act like a grumpy conservative lawyer. Instead your brain will be the suspect. You, on the other hand, will play the role of a juror whose task is to find the truth. The definition of truth in a courtroom, remember, is "**the truth, the whole truth, and nothing but the truth.**"

According to that definition, I'll dare say that *none* of the endless chatter in our head is ever "entirely" true. *Yes—none!* "That is a big statement, Mo. Prove it," you say. I will.

First, I would like to call an expert to the stand to explain the seven blind spots in detail.

Filters

The picture we see of the world is always incomplete because our brain omits parts of the truth in order to focus on what it deems a priority. What we perceive is mostly filtered, leaving us only a tiny sliver of the truth.

The world throws information at you every second of every day. Through your senses, you have the ability to observe every variable. The temperature of the room, the brightness of the light, the background sounds, the movement of a fly, the words of a friend, and millions of other stimuli. Most of this information isn't relevant to every decision you have to make at any given moment. And your brainpower, although unmatched by the largest supercomputer we've ever invented, is still limited. As a result, your brain optimizes its resources carefully by filtering out details that are irrelevant for the situation at hand. This enables it to focus on the essential data that is critical to the decision it needs to make.

When you try to cross the street, your vision brings you information about the cars approaching and their anticipated speed and direction. Your brain calculates the distance you need to cross. With instinctive knowledge of trigonometry and dynamics, it assesses for a possible collision point. It instructs your eyes to focus and to check for any red lights or traffic signs and sharpens your hearing to detect the horns of drivers trying to alert you. It coordinates your muscle movements to look left and right just as an extra precaution—then you decide to step forward.

We do all of this in a split second. But if you tried to program such functionality into a robot, you'd quickly realize just how hard it is to achieve. Obstacle avoidance requires a very complex spatial calculation mixed with a highly advanced operation of muscle coordination. This demands a lot of processing power. And because even a tiny mistake could mean your life, your brain takes this task very seriously and gives it undivided attention. So what does it do? It filters.

While crossing the street, you'll not pay any attention at all to the smells surrounding you. You'll be listening for horns and sirens but muting most other irrelevant sounds, such as the chirping of the birds on the tree around the corner and the crying of a baby far behind you. If the approaching cars are moving fast enough to grasp your full attention, even a pretty girl in a short skirt or Brad Pitt crossing from the other side will pass unnoticed. Yes, filtering is that effective.

> **After reading the**
> **the whole sentence you will**
> **become aware that the**
> **the human brain often**
> **does not inform you that the**
> **the word "the" has been**
> **repeated twice and filtered**
> **out every time.**

Daniel Simons and Christopher Chabris designed the Selective Awareness Test to demonstrate how this kind of filtering works. They asked participants to watch a short video showing two teams dressed in either black or white T-shirts passing a basketball between them. Participants were given the task of counting the number of passes made by just the white team, which isn't rocket science. Human brains, however, take this kind of thing very seriously, and they focus like demons. Try the test yourself before reading further; you can simply search for "selective awareness test" on YouTube.

Most observers report the number of passes accurately. But when asked about the gorilla (Yes, a man dressed in a gorilla costume walks across the screen waving his arms halfway through the video), more than half of them respond "What gorilla?"[9]

You experience this kind of filtering when you enter a movie theater. At first you notice the empty seats, the people, the smell of popcorn, and the annoying light of the exit sign. Once the movie starts to draw you in, though, you filter out all that extraneous stuff and fully tune in to the movie. You become oblivious to your surroundings, and if the movie is good enough, you don't even notice the passage of time.

Filters are used to reduce our pain or emotional reactions when what we're facing goes beyond our ability to cope. Extreme pain is filtered in the case of a broken bone, for example, in order to enable the brain to focus on getting help. In the case of the loss of a loved one, the first stage of the process of grief famously starts with denial, which in itself is a mechanism used by the brain to manage the distress by filtering out the event and dismissing the loss as if it never even happened.

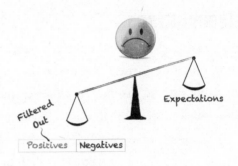

When we push filtering to the extreme, our ability to focus works against us. We sometimes obsess about one thing that makes us unhappy, and we filter out any positive signals that could change our frame of mind. As we do, we let in more and more signals that match our filter and confirm our reasons for being miserable. When you filter out the truth, your inputs into your Happiness Equation become distorted. You suffer, not because life did not give you what you expected but because you failed to notice what life had actually given you.

If you do the math and realize how much you actually filter out, the result will shock you. At any point in time what you filter out is orders of magnitude more than what you let it. Try it yourself. Put the book down for a minute and look around you. Notice the magnitude of intricate detail you've missed—filtered out—while focusing on the pages of this book. Count the number of objects you're starting to notice, the colors, smells, and sounds that you omitted until you eliminated your filters. Now quickly calculate how much of the truth that represents and you'll realize that because of its filters:

Remember !

→ **The story your brain tells you is *always* incomplete.**

Assumptions

To make its decisions, the brain needs a coherent, comprehensible set of information. After filtering out the majority of the truth, the brain then goes on to assume any information that seems to be missing. Reading a misspelled word is a clear demonstration of such abilities.

> -t's n-t h-rd f-r th- br--n t-
> -ss-m- th- m-ss-ng v--ls -nd
> r--d th-s s-nt-nc-

Assumptions distort the truth even at the physical level of visual perception. The term I'm using here, *blind spot*, is usually used when someone fails to observe something important. But in anatomical terms, blind spots are the parts of your visual field where you can't actually see because your retina lacks the necessary cells where it connects to the optical nerve. With no cells to detect light, a part of the field of vision isn't perceived; you would see this as a black spot were it not for your brain's ability to make assumptions. The brain interpolates the blind spot based on surrounding detail and information from the other eye when available, so the blind spot is replaced with the image it is likely to contain. While the resulting picture looks perfect, it's not entirely true, as parts of it are generated by your brain.

Attempting to assume what's missing is benign, perhaps, but changing what you actually see in order to match the brain's own expectation is taking it a bit too far. A famous experiment by Edward Adelson of MIT demonstrates the way our brains do that by using the image of a checkerboard. Which of the squares (A) or (B) is darker? The answer is clear, isn't it? (A) is obviously darker than (B).

But that is the wrong answer! Take a look at the same image with all but the squares in question faded (you can do that yourself by covering parts of the image). Which square is darker now? When the image is viewed this way you'll see the truth. The shade of the cylinder darkens the white square (B) enough to match the actual shade of the well-lit square (A). But due to our familiarity with the chessboard pattern, our brain assumes what the appropriate shade of (B) "should" be and uses that as the view you, in fact, end up seeing.

The most incredible part of this brain trick is that even as you now know the truth—that both squares are exactly the same shade—if you take another look at the first image your stubborn brain would still render the "assumed" untrue image. Try it!

Now take that concept from vision and apply it to thoughts in general, and you'll realize that we make assumptions all day long. We assume that a man is stronger than a woman, that gray hair signals wisdom, that rich means successful, that skin color . . . Don't get me started. We fall prey to these biased assumptions all the time.

And in our modern social circles, assumptions multiply and bend our perceptions of reality way out of shape. Here's a world where the threats are no longer tigers but come in the form of nasty coworkers, unfaithful lovers, and economic crises. Such events are so complex that it's beyond anyone's capability to grasp the infinite and intricate details that make them up. As we fill in the blanks for such complex scenarios, events are morphed into elaborate stories that edit a significant part of the truth. If the reality presented to your brain is "My boss missed her target last quarter," you may assume that she's under pressure and therefore she's afraid that your achievements will make her replaceable. This may lead you to assume that she's out to get you, and therefore you'll assume, still with no foundation, that she is trying to make you fail. In conclusion, you'll assume your boss is your enemy and behave accordingly.

Alternatively, if you adopted a more positive mind-set, you would observe the same fact, "My boss missed her quota last quarter," and build a very different story. You may assume that it's necessary for your team to make it this quarter, and therefore you'll further assume that your boss will do everything she can to make you succeed. In conclusion, you'll assume your boss is your ally and behave accordingly.

Both are plausible scenarios, but neither of them is indisputably true. They're both just a sequence of assumptions that need to be verified as more facts become available.

Dozens of similar situations happen every day. To keep up, our brain makes more and more, quicker and quicker assumptions—and moves on. This often results in stories that contain more assumptions than actual facts.

Unfortunately, since the grumpy brain is designed to prioritize your survival, more often than not it will make up a grumpy story that's bound to make you sad or worried. But remember, those stories are not true because:

Remember !

→ **An assumption is nothing more than a brain-generated story. It's not the truth!**

Predictions

Our brain makes assumptions to fill in the gaps. And what's the biggest gap? The future. We know nothing about what's coming. The future may swing a million different ways. Nothing about it is certain, but this won't stop our brains. They shamelessly fill the gap.

Our brain can connect two or more points of data from the past and present to establish some kind of trend, and then project fictional future scenarios based only on extrapolation. For example, if your best friend's boyfriend cheated on her, and that hot guy in the morning soap opera cheated on *his* girlfriend, then your brain may connect the two dots to establish a possible trend: that *all* men are cheaters. It then extrapolates that trend forward to expect that *your* boyfriend is bound to cheat on you. Your predictive engine then starts to build a story: you remember that last week your boyfriend said "Hi" to your neighbor, the same one who flirted with him once a year ago. *What a cheat!* You can see where this is going. You consider your prediction to be true beyond any doubt, and you forecast the ending of the story. Accurate? Not even close—but at least the story is complete. And that's when it becomes more interesting.

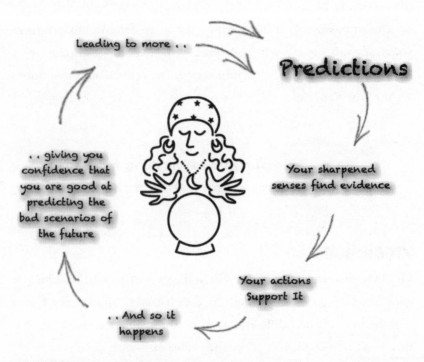

When you predict that your boyfriend is going to cheat, you start acting as if he already has, and then he just might. If he does, you'll say, "You see? I told you this was bound to happen. Victory, my prediction came true!" But was it a prediction or a cause? And how often do our fears of the future help create the reality that we fear? We'll never know.

I just know this:

Remember!
→ **Predicting something will happen often lays the path to make it happen.**

We extrapolate, forecast, and predict all the time, and as our forecasts change our behavior, it frequently delivers what we predicted. The more this happens, the more we start to believe that our predictions are the truth. Your clever brain no longer presents predictions as *possible* future scenarios that may or may not happen. Instead, it presents the future scenario as a fact that must be considered in your assessment of current events—and your hope for happiness goes out the window.

But here's the fact:

Remember!
→ **Your predictions are nothing more than brain-generated future possibilities. They have not happened. They're not the truth!**

Memories

Our brains then look back to intermix our perceptions of current events with memories from our past. At work, for example, we assume that something won't work just because we've tried it before and failed. Such bias ignores the possibility that the circumstances of the previous attempt may have differed drastically. Blinding the current preposition with memories of a previous struggle leads to decisions that are not entirely based on the reality of the current situation.

We all do this kind of blending. In our personal lives, we often create impressions of a person we're meeting for the first time from a memory of someone who looks similar. We mix our memories with our current reality to create an augmented view that's colored by the past.

If you mixed a gallon of pure water with one tiny drop of ink, the resulting liquid, diluted though it may be, would no longer be pure. Memories are like that drop of ink. Mixing them into the current reality creates a richer, more familiar and augmented story, but one that is no longer a pure reflection of the truth. And then it gets worse.

If you mixed an invisible contaminated substance—say, a virus— with that gallon of water, the risk you face is somewhat contained. But if you then threw that gallon in the main supply for your water, you can rest assured every drop of your water will be contaminated for a long time to come. And that, unfortunately, is what we do when mixing memories with present realities.

We view memories as archives of past events—of what has actually happened. But in reality, memories are nothing more than descriptions of what we *think* happened. And because what we think is always distorted by our brain's blind spots, it's often *not* true. We augment those stories from the past, inaccurate as they may be, with the pure reality of the current events, producing a dangerous mix and consider that to be the truth.

You and your girlfriend may go to a beautiful place for the first time but end up fighting, so your memory of the place gets logged as sad. When you go the next time, your perception of the place is contaminated by that memory and your assessment of it becomes biased to sadness. Here's your contaminated gallon. Then things get even worse. You log the new experience—made up of a current reality augmented with a sad past memory—as a new sad memory ready to be recycled into the next story. The margin of error in your perception is multiplied with every repetition of the cycle of blending past and present. This endless loop progressively disfigures your perception in consecutive cycles and sways you further and further away from the truth.

Don't contaminate your perception of current realities.

Remember!

➤ **Your memories are nothing more than a record of what you think happened. They're often not the truth!**

Labels

Memories augment the truth with a series of events from the past. Labels also come from the past, but are more potent. They take the form of a simple tag without the memory of a specific event attached. Our brains judge and label everything, then turn the results of such analysis into short codes by removing context and details. They use those labels to enable prompt decisions, but in so doing they sacrifice accuracy.

A Middle Eastern man with a long black beard is automatically labeled a terrorist. A rainy gray day is labeled miserable, and an exotic-looking car is labeled fast. Such labels are a result of repeated associations. If people who look a certain way repeatedly appear on your regular news network along with an anxious anchor repeating the word *terrorist*, then your brain will comfortably associate one with the other. This allows your brain to be a lot faster. It doesn't need to redo the whole analysis and association; instead, with one quick access to its database, it can make split-second decisions based on the label already on hand.

It might be useful to take a look around the next time you're in a crowded place and notice how many judgments you pass in the form of labels. She's *short*. He's *scary*. It's *too bright*. That's *too expensive*. What a *bargain*. All of those condemn something or someone to a category—praise or criticism—as they stop you from looking more closely to observe the bare reality.

Labeling is so instinctive even monkeys do it. In a famous experiment, several monkeys were placed in a large cage where a bunch of bananas hung from the top of a ladder. When a monkey spotted the bananas and began to climb the ladder to get them, the researcher sprayed the monkey with a stream of cold water. Then he proceeded to spray each of the other monkeys too. The monkey on the ladder scrambled off, while all the others sat on the floor, wet, cold, and very unhappy. Soon, however, the temptation of the bananas overwhelmed one of them again and it began to climb the ladder. And again the researcher sprayed all the monkeys with cold water. It didn't take long for the troop to learn the score; when the next daring monkey attempted to approach the ladder, the others quickly pulled it off and beat it down to avoid being sprayed. Those monkeys associated the act of climbing the ladder with an unpleasant experience and created a label. Even when the spraying stopped, they still avoided reaching for the bananas, because to them, the association was clear: ladder=cold water. They missed out because labels inherently cover up an interesting part of reality.

Labels preempt further analysis, which causes us to miss out on the context. When climbing triggered cold water, it made sense to avoid the ladder, but when the context changed, the label only kept the monkeys needlessly hungry.

And we miss out on so much reality because the context of labels varies across cultural backgrounds, age groups, and a million other variables. In the West, for example, a thin tanned woman is assumed to be rich and gets labeled as such. Those features seem to indicate that she has the leisure time to take care of her figure and spend time in the sun. In contrast, in many parts of Africa, rich women tend to have fuller figures and lighter complexions; such features indicate that they have plenty to eat and no need to spend time working in the sun. A thin, dark-skinned African woman would likely be labeled poor.

Anything that inhibits our ability to be in touch with the truth also inhibits our ability to solve for happy. When we label, we collapse the diverse possibilities of how events actually are into an approximation at best—a snap judgment that might not reflect the truth. And whenever we use false inputs into our Happiness Equation, we fail to solve it correctly and we suffer. Apart from that, labeling takes away the pleasure of living a full life by rendering it in a handful of colors and names when, in reality, the world is an infinitely diverse patchwork. When we label, we dampen the richness that life has to offer.

I can deeply relate to labeling because it was always the blind spot that Ali disliked most. In his university admission essay he wrote about how, as a teenager with super cool dreadlocks, he suffered as he traveled between East and West. In the West, he was labeled because of his name, race, and religion, while in the East he was labeled because of his culturally unacceptable looks. He wrote, "How could anyone get to know the truth about who I am, until they get through my race and dreadlocks?" Labeling never truly changed him, though. When he was fourteen, the father of the girl he deeply loved asked him to stay away from his daughter because of his Eastern origin. Honest as he was, he stopped calling or texting her for more than eighteen months, until that honesty made her father realize how he had labeled Ali. Eventually the father changed his mind and allowed them to be together. Ali went on to live true to himself regardless of how he often was labeled. When he left our world, his English teacher wrote a post describing him as "the guy who was unapologetic about following his own rhythm." I, on the other hand, remember him as the guy who taught me to see the truth in so many ways, not the least of which was this:

Remember !

➤ **In the absence of context, labels very often cover up the truth.**

Emotions

Emotions make us human, but when we blend them with logic, they can impair our judgment. While most of our decisions are (ideally) driven by logic, most of our actions are actually driven by emotions. We work hard because of ambition, love, and desire. We hide away because of fear or shyness. Even seemingly cold-hearted politicians and executives are motivated to act by their emotions of pride, anxiety, and fear. Our emotions are ever-present because they represent a critical component of our survival machine.

If that tiger that scared our species during the caveman years eventually does show up, an extreme emotion—panic—would overwhelm you. Your brain will snap into full alertness, realizing there's no time for chitchat. It will suspend your normal thought process and dedicate all your physical resources to the situation at hand. Adrenaline will flood your body—and that's when the miracle will happen. *Zoom*, you sprint out of harm or pounce on the tiger to slit its throat with one confident strike. In order to unlock those kinds of superpowers, emotions have to take over.

Today, despite the absence of physical threats, our modern brains still won't allow themselves to sit idle. They keep busy by engaging emotions around imaginary threats. Events that wouldn't have crossed the mind of our ancestor caveman now seem to be central to our emotional well-being. If you could have asked the caveman where his "income" was going to come from, he would have looked puzzled and said, "Tomorrow, we go hunting." And if there is nothing to catch? "Then we go out the next day." And what will happen when you get old and can't hunt anymore? "The tribe goes hunting." How about your health insurance, children's tuition fund, and retirement planning? "Huh???"

Compare our modern lifestyle with the past and you'll understand why life has become so stressful. Although harsher, life then was much simpler; that's because the emotions of our ancestors were more in harmony with the norms of the animal kingdom. Antelope, like us, can experience fear. When a tiger becomes an imminent threat, the antelope will rapidly move from calm to fear to panic. Its heart will start to pump faster and a miracle reaction will happen: it'll run like the wind. Throughout the chase the antelope turns at sharp angles and jumps in and out of gullies, outpacing the mighty tiger. A few minutes later, it manages to maneuver away from danger, and then, just as suddenly, it goes back to its calm state and stops to eat some fresh grass as if nothing had happened. The tiger, on the other hand, won't linger if the prey escapes. It won't blame itself for being too slow on that last left turn, and it won't feel ashamed in front of the other tigers. Once the prey escapes, the tiger also goes back to its calm state and sits quietly, unbothered by the flies on its face. Inspiring!

We modern humans behave differently. We're often engaged in some kind of emotion and, frequently, several—sometimes contradictory ones—at the same time. Many of those emotions keep us in a state of unhappiness. Yet we keep them active—sometimes for a lifetime—though we don't always admit to their influence.

That endless flood of human emotions raises the question as to whether we are as rational as we might think. In one of Plato's dialogues, Phaedrus describes reason as a charioteer who holds the raging emotions of his horses in place. This image reflects the Western orientation toward our distrust of emotion, which has helped to build a culture that worships at the altar of rationality. We're trained, especially in professional relationships, to prioritize logic, tamp down our emotions, and keep them hidden when they arise. The irony is that our emotions are still in full control. The reality we cover up is that we tend to make our decisions based on emotions first, and then gather the data that support those decisions. If you really want to buy a new TV, you may decide within a few seconds that it's a real bargain, then start looking for reasons to back that up. As you look for the good side of the deal presented, you tend to overlook the downsides and accordingly end up carrying the TV home with you. The opposite is also true. If you belong to a specific political party, you'll decide to dislike the speech of a candidate from a rival party even before she speaks. Then, when she does, you'll gather evidence as to why it was a bad speech. When you consider all this, you realize that Plato's horses are in full control. Perhaps it's time to acknowledge this simple truth so that you can make the horses take you where you really need to go.

Remember! **We're not as rational as we think. Our perception of the truth is often distracted by our irrational emotions.**

Exaggeration

You have to admire the amazing persistence of our brain. Its firmest principle is *You can never be too careful.* If the truth isn't enough to convince us to take action and run for cover, our brains will exaggerate perception to grab our attention.

And exaggeration works. It totally grips you—and every other species on the planet. It's not difficult to teach a research lab rat to distinguish a rectangle from a square. All you need to do is give it cheese every time it picks the rectangle. The association reinforces the behavior, and soon it will select the rectangle every time. Once the rat develops this preference, you can start to notice a feature called "peak shift," a preference for "exaggerated"—longer, skinnier—rectangles. What the rodent has learned to recognize is not a particular rectangle but rather rectangularity itself: the more rectangular a shape is, the more attention it will get. The rat's strongest reactions align with the most exaggerated deviations from the norm.[10]

This feature makes female peacocks prefer males with large tails and enables the strongest male lion or gorilla to get all the females. And, naturally, peak shifts are even more true for our more sophisticated species. Women, looking for a fit father for their offspring, will seek a mate with good genes and stability. They are attracted to strong visible physical strength, indicating good genes, but also to apparent wealth, big-shot career, and success. The more exaggerated those elements are, the stronger the attraction. Hence the success of industries that depend on branding as a show of wealth and success. Men, on the other hand, are instinctively attracted to women with exaggerated body proportions, indicating fertility. They are drawn to bigger, well, you know what I mean—features. Hence the massive success of the plastic surgery industry.

But none of those exaggerations are true characteristics. They could be nothing more than an inflated appearance and might not in any way be accompanied with real wealth or fertility. The exaggeration deceives us, but more important, when the negative is exaggerated, it causes us to suffer.

When a negative event is exaggerated, we worry about it even if it's statistically unlikely to harm us. Airline crashes, shark attacks, or terrorism occupy our mind, while everyday dangers that kill thousands pass unnoticed. Princeton professor and Nobel Prize winner Daniel Kahneman calls this "the availability heuristic": if you think of an incident in which a risk is confirmed, you—your brain—will exaggerate its likelihood. "Somehow the probability of an accident increases [in your mind] after you see a car turned over on the side of the road," says Kahneman.[11]

On the other hand, events that are not exaggerated are ignored despite their true magnitude. Consider events that get little coverage in the media. Paul Slovic, a psychology professor at the University of Oregon, says, "With 9/11 we lost 3,000 people in one day, but during 1994 in Rwanda 800,000 people were killed in 100 days—that's 8,000 a day for 100 days—and the world didn't react at all."[12]

By broadcasting their exaggerated views inside our heads, our brains use our peak shifts and availability heuristic tendencies to grab our attention. And as they succeed in keeping us focused, the price we end up paying is unnecessary suffering. We read too much into what a friend said, exaggerate the threat of being unemployed, and magnify every fear and worry. In a modern world that's too noisy, the exaggeration goes overboard, inflating a sizable proportion of what our brains present as truth.

Exaggeration in all its forms inflates our expectations and destroys our satisfaction with life, regardless of how pleasant life may actually be. An exaggerated view is bound to make you unhappy. More important, it's not even accurate. Exaggeration adds layers of fiction to the reality, and that's a lie.

Remember!

What's more than the truth is less than true.

Closing Remarks

In criminal law a convict is innocent until proven guilty. But that is not the case with our brains. **In "brain law" your brain is guilty until proven innocent!**

Shawn Achor, a professor of positive psychology at Harvard, says, "What we're finding is that it's not necessarily the reality that shapes us but the lens through which your brain views the world that shapes your reality. If I know everything about your external world, I can only predict 10 percent of your long-term happiness. Ninety percent of your long-term happiness is predicted not by the external world but by the way your brain processes the world."[13]

Life As Seen Through the Lens of Our Blind Spots

The mismatch between events and expectations in your Happiness Equation is often a matter of feeding your thoughts with wrong information, not a matter of what life actually presents to you. The events we factor in are distorted and our expectations are inflated. Both sides of the equation are messed up—but two wrongs don't make a right. As a matter of fact, two wrongs make things *very* wrong! Enough with the needless suffering.

Teach your brain to tell the truth, the whole truth, and nothing but the truth.

Debugging Your Brain's Code

Like every software program, our brains follow hard-coded routines. When a program has a bug, it performs the same suboptimal task over and over every time the program is run. The tough part of programming is to find the bug. But once you pinpoint the problem, it's generally an easy fix. The same is true with our brains.

Filtering is an easy bug to spot if you remember that nothing you perceive is ever complete. There's always more to discover. When you notice that your brain is summing up a complex set of events in a short sentence or is obsessively looping around one specific thought, ask it, "What part of the story are you filtering out, Brain? Is there something more I should know before I make a decision?" Your brain is just a tool. Ask those questions and it will respond. "Oh, I forgot to tell you this," it'll say. Ask again and again until you see as much of the truth as you need for an objective view.

Seeing the bare truth without assumptions starts with parsing out what you can verify with sensory perception. If you haven't sensed it, then you're making it up. An easy way to spot the assumptions is to understand that true events in our life are described with verbs such as *I saw, I heard, I was informed,* and *I noticed,* while the stories we make up use verbs such as *I guess, I feel, I assume, I think,* and even *I'm sure.*

This kind of linguistic alertness also helps you spot the memories bug. Those show up in your head in the past tense. Thoughts like *This is how it used to be,* and *I knew him when,* and *Those were the days* are examples of thoughts anchored in the past.

Predictions, on the other hand, are associated with future tenses. Consider those for what they are: unsubstantiated predictions. If it didn't happen yet, it's just a forecast, not the truth, regardless of how convinced you are that it will happen.

Labels normally come in the form of short, snappy, yet confident judgments: *He's an idiot* or *This place is a dump*. They even pop up as single words: *pretty, scary, silly,* and a million other words of praise or accusation. They seem to describe a complex topic in one word for quick decisions. When you see them, you've spotted a judgment you should scrutinize and a label you should remove.

Removing all emotion from your thought process is neither possible nor desirable. But as you observe the dialogue in your head, look for signs of emotion tinting your perception. Verbs like *I feel, I love,* and *I hate* and phrases of heightened emotional engagement such as *She's a pain, He's unbearable,* and *They're annoying* are indicators of an emotion weighing in. Once you spot them it'll be easier to parse out the truth.

Exaggerated thoughts are normally marked with blanket statements that tend to generalize: *huge, tiny, never, always,* and so on. When you see or hear any of those words, pay attention. You're brain's blowing things out of proportion.

Remember that the truth, unobstructed by the blind spots, normally sounds dry. "She's drop-dead gorgeous," for example, becomes "She has symmetrical features, large blue eyes, long hair, and a nice complexion." While that may not get you very far as a compliment on a date, it'll take you much further when solving the Happiness Equation.

Troubleshooting each of the blind spots separately will enable you to take big steps forward. But remember, those features have been with us for millennia, and they're not going away soon. Just as my old car was at the mercy of one mechanical problem after another, our brains will always be influenced by one blind spot or another. When that car kept breaking down every other week, I wished there was a magic fix that I could do once and for all. That did not exist for my car, but it does exist for our brains.

It's all summed up in one simple question: Is it true?

Is It True?

In her book *Loving What Is*, Byron Katie uses this question as the pillar of her model. Katie developed a method of self-inquiry that she calls The Work, a system for discarding the stories we tell ourselves and replacing them with the truth ("what is").

Let's start with a simple example:

> *My teenage daughter is a pain in the neck.*
> Is that true? Does she make your neck hurt?
> *Oh, you know what I mean. She's impossible to deal with.*
> Is that true? How can it be impossible if you've been dealing with her so far?
> *I'm just saying, she's always rude.*
> Is that true? Always? Like every single second of every day?
> *No, not always, but she shouldn't be rude anyway.*
> Is that true? So teenage girls are not, sometimes, rude? Where have you been living?

Keep asking the question "Is it true?" as many times as you need until you realize how ridiculous the statements our brain offers us really are. Keep questioning until you end up with a description of the event that's a factual narrative, a story that attaches nothing more to it than the truth: My daughter's recent behavior indicates that she might be a bit irritated. That's the simple truth.

Practice this technique as often as you can. I'm confident that your brain will give you an endless supply of practice material.

The Truths

Most of the time the only thing wrong with our lives is the way we think about them. When you see the world for what it really is, you solve the Happiness Equation correctly. And the more you do that, the more you will notice how often the events of your life—seen correctly—almost always meet your expectations when set realistically.

This will eventually drive you to ask yourself this question: If the reality of life mostly meets realistic expectations, why should we bother with the Happiness Equation at all? That, my friend, is a *great* question.

We'll answer it at

You
Are
Here

6
Grand
Illusions

7
Blind
Spots

5
Ultimate
Truth

Part Four

Ultimate Truth

5 ultimate truths are all you need to know to realize that life always behaves as expected. Those truths will solve your Happiness Equation once and for all. Events, even the harsh ones, always meet the expectations of a wise mind that knows how life really behaves, not how it wishes life would behave. None of the twists and turns of your unfolding life will matter anymore because you expect them and know exactly how to deal with them. If you anchor yourself in the truth, you move above thought and into the bliss of peace, where nothing shakes your happiness. You move above a state of happiness that is conditional on external events into a state of permanent joy.

Before I get the courage to discuss with you what I claim to be the truth, I need to make a few things clear.

The truth will set you free. I know this is a cliché, but it's true. Loss, lack, and pain; love, growth, and inspiration—they're all part of life. We each get a share, and while they usually show up when they're least expected, it's hard to imagine a life without them—each of them.

When Ali left our world, losing a loved one became a topic central to my life. Friends kindly approached me and shared their own stories of loss. Many of the stories I heard were even more shocking than mine. It surprised me how many people have had to go through excruciating pain yet walk among us with their pain unnoticed. I started to wonder if there was anyone who didn't have to endure such tragedy. Since I started to visit Ali's grave, hundreds of others have come to join him as his neighbors. Workers dig new graves at a steady, predictable pace. I see families and friends visit. They often follow a pattern. The chaos of the first encounter is followed by months of deep grief. Visitors cry and despair. They visit often and they stay long. I sit still next to Ali and observe, and one day one of those visitors smiles. It normally happens after a few months have passed. They might tell a story to their departed and say they miss them. As the time ticks along they visit less often, leaving those parts of the graveyard deserted while the previously deserted parts become busy with new graves. This made me wonder if even death, in a way, is expected. Death surely is unwelcome, intrusive, painful, and untimely, but who can say it's *unexpected*. Death is very real. It should be expected.

So is the nature of every truth. We reject it and wish it wasn't true, but it overpowers us. We dwell on the past and worry about the future, while we can influence nothing but the present because *now* is real. We try to stay in control and make our life predictable, but eventually we get taken over by black swans and butterflies because *change* is real. We resist and disbelieve to no avail because:

> Remember !
> ➜ **Every truth happens exactly as expected, even when you least expect it.**

And that's a good thing because when your reality matches your expectations, your Happiness Equation is solved. Life, with all its harshness, fails to shock you and you finally find peace.

I wouldn't have been able to survive Ali's loss had it not been for my establishing a lifestyle built on accepting death. This is easier said than done, I admit. Like a master archer, you need to not only see your target but, more important, see nothing else *but* your target. Your target is the truth. Seek it relentlessly but recall Gloria Steinem's warning:

> Remember !
> ➜ **The truth will set you free, but first it will piss you off.**

We've come so far together, hopefully as good friends, and I'd like to keep it that way. So here's my disclaimer about what I *claim* to be five ultimate truths: Making a claim that something is true, let alone the ultimate truth, is at extreme odds with the Illusion of Knowledge. Nothing is indisputably true. These are *my* five truths. They helped *me* find joy and survive the tragedy of losing Ali. Every event that I faced in *my* life, harsh or pleasant, seen through the lens of these truths seemed expected. Events, though many battered me, met my expectations, and my Happiness Equation has therefore remained permanently solved.

Some of my truths, I expect, you will agree with. *Now, change,* and *death* are real. The others, namely *love* and the *grand design*, might be controversial. Like many, I rejected those last two truths for years, but then I found answers in logic and mathematics that compelled me to change my view. All I ask is that you read my logic and be open to an alternative point of view. If you continue to disagree, that's perfectly okay. You can seek your own truths. It doesn't matter what you find as long as you treat *your* truths as *your own* signposts to find *your own* path to joy.

The truth—always—is just a single dot on a long line of infinite possibilities, of which every other point is an illusion. That's why the truth is hard to find. But here's an easy test to help you: if you find that a certain concept leads to your suffering, then perhaps you should doubt its validity as truth. We're not here to suffer but rather, as Arianna Huffington has said, "to be whittled and sandpapered down until what's left is who we truly are."[1]

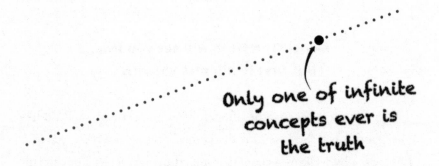

Only one of infinite concepts ever is the truth

When you're searching, some concepts will be easy to pinpoint as an illusion, while others will shine as obvious truths. There are, however, points on the perimeter of the truth where it's hard to prove either way. This is when you need to make a crucial choice and follow **Golden Rule for Happiness**: *Choose to believe in the side that makes you happy.* That side is more likely closer to the truth.

When I find it hard to prove for or against a specific view, I choose to believe in the side that makes me happy. Choosing the side that makes me suffer with no evidence to prove my view is, well, not a very smart thing to do.

Remember!

→ **When nothing is certain—and nothing ever is—choose to be happy.**

This rule will become our backbone when it comes to some of the more debatable truths. For now, let's start with one that's undisputed: Now is Real.

Chapter Ten

Right Here, Right Now

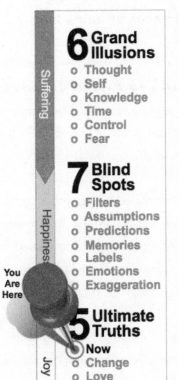

All of life is *here and now*. So why do most of us live *there and then* instead? Why do we live in our head, outside the present moment, fully absorbed in the Illusion of Thought, unaware of the beauty of the unfolding life all around us? Why do we let our absence from the present cause us so much suffering? Because that's how we were trained to be.

Matt Killingsworth of Trackyourhappiness.org ran a study with more than fifteen thousand participants who reported, sometimes minute by minute, how they felt and what they were doing at the time. The study collected more than 650,000 reports and presented a profound discovery: regardless of what people were doing at any given time, they were noticeably happier when they were fully present. It didn't even matter what they were thinking about. It didn't matter if it was a pleasant, a neutral, or an unpleasant thought; when they were focused outside the present, people were less happy. Period.[1]

Matt explains, "If mind wandering was like a slot machine, you would have the chance to lose $50, $20 or $1." Wandering to unpleasant, neutral, or pleasant thoughts makes you lose any which way. Better not to play at all.

Remember!

➤ **Being fully aware of the present moment considerably increases your chances of being happy.**

What Is Awareness?

Awareness—a sense of knowledge or perception of a situation—is our ability to grasp the world at any given moment. Presence—the state of existing, occurring, or being attentive—is what enables this awareness.

Pretend that I held a talk to discuss happiness. Unless you actually manage to be "physically present" in a setting that allows you to hear what I say, you would not be "aware" of what I was discussing. (I'll revisit this statement a bit later, so don't take it for granted just yet.)

Mere physical presence, however, isn't enough. You could be sitting there feeling bored. The sound waves could be hitting your ears so you hear noises, but you wouldn't be *aware* of what I was discussing. Without the intent to be aware, there's only *reception* without awareness—a state all too familiar in our modern world.

Sometimes we even fall below that state into a state of *rejection*. If you plugged your ears during my boring talk, the sound waves would be attempting to hit your eardrums all the same, but you wouldn't be letting in even the noises.

Awareness starts to emerge when you pay attention. In this state you're interested in what's happening. The interest tunes you in, and you understand the waves hitting your eardrums as words and concepts. This is *perception*.

The more emphasis you put on your intention to be aware, the more you pay attention and the more you perceive. If the topic deeply interests you and a member of the audience stands up in the farthest corner of the room to ask a question, you would turn your head and sharpen your senses to hear. You don't want to miss a thing. This is *awareness—when you're fully immersed in the moment*, fully aware of what's happening. For me, this is the stage where I start to feel alive.

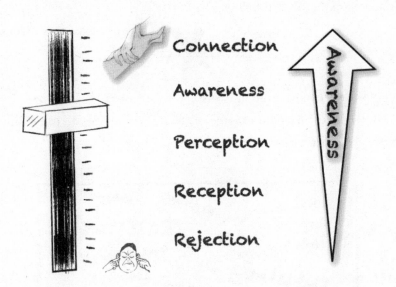

Sometimes you become so plugged in to what is going on that you start to notice signals that no one else does. For example, from the facial expression of the person sitting next to you and his body language you might sense that he disagrees with what's being said. You become hyper-tuned-in, trying to collect all the data around you, so you even surprise yourself at how much you become aware of. This is what I call *connection*.

Remember!

➤ **Awareness isn't an on/off switch. It's a dimmer switch. When you choose to crank it up, you become more aware.**

Maintaining awareness seems to be hard for some of us, especially in our increasingly distracting world. To understand why, let's start with a question: What are some of the things you can do to become fully aware? I'll give you a few minutes to think about that. Don't cheat and look up the answers below.

Now let's compare our answers.

In the box, I have listed some of the things you can do to be fully aware. Yes, sorry, it was a trick question.

Making the Connection

Remember the Full Awareness Test from chapter 6, where you close your eyes, then open them for a couple of seconds and just try to grasp the room around you? Remember how much you could take in within just a couple of seconds? Did you have to do anything to see all that's around you? Was there any *doing* involved? No, there was none. The second you open your eyes with the intention of being aware, you are aware. Your sense of awareness is always ready to receive. All you can do is cover it up.

You might feel resentment for a speaker. That emotion consumes you and takes your awareness away. You start to draw little hearts, focusing on what you draw, or you direct your thoughts to another time or another place. Doing removes your intention to pay attention. If you just stop doing, you will default to being. And being is the only state in which you're fully aware.

The Full Awareness Test simply gives you two seconds to *stop doing*. Those two seconds are all you need to find your true self and become fully aware.

Very Important!

➤ **You don't need to do anything to be aware. Your default status is awareness. To reach it you need to stop doing!**

But living demands that we alternate between the states of being and doing. Some of us spend more in one than the other. Most of us *do* more than *be*. It's what the modern world expects of us. We wake up every morning and rush through a life that is totally engaged in doing. This fast-paced, immersive lifestyle is the opposite of our default nature as humans. It's like living underwater while wearing heavy

shoes. Everything around you is hazy, unfamiliar, and heavy. It's hard to move or function naturally. You feel fatigued as you push against the viscosity of the water. You feel the pressure of the depth and the lack of oxygen to breathe. Your eyes burn with the salty water, but you keep trying to find your way, totally exhausted and performing below your best. As harsh as this definition sounds, it's very close to how we go through life while being totally unaware.

Does this sound familiar? It sure does to me.

All that doing and thinking of modern life leaves no space to let awareness in. By removing the clutter, we become present, pay attention, and start to receive. You cannot fill a glass that is already full. You have to throw away the stale water to let fresh water in.

You don't *do* aware. You *be* aware.

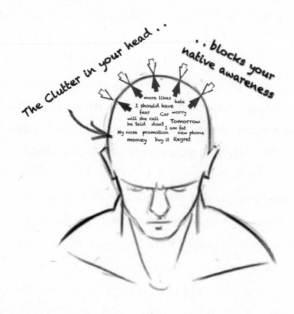

Remember! → **Stop doing and just be.**

As you do, you'll find it fascinating to realize that often, doing is not the only way to achieve progress and results. Sometimes you can get ahead just by being—a concept that is at extreme odds with modern Western culture.

 The Taoist tradition captures this in a concept called *wu wei*, which translates as "nondoing." A metaphor often cited in this philosophy relates to farming. If your intention is to grow a plant, do what you should do. Give it sunlight, fertilizer, and water. Having done that, begin the nondoing by leaving the plant alone to grow on its own. Once the conditions for growing the plant are fulfilled, more doing results in more harm than good. The wise farmer knows that the best possible progress is achieved by inaction. Doing nothing is the best thing to do.

Remember! → **The result of awareness is a net positive especially when there's nothing to do.**

The Extents of Your Awareness

Many meditation techniques identify four corners as the extent of our "awareness space." Direct your attention to any corner and you will find an endless spectrum of subjects worthy of your full awareness. Those four corners are:

The world outside. Through sensory input you can grasp the world around you. Perceive the sights, sounds, smells, tastes, and feelings of touch.

Inside your body. By directing your attention inward, you can become aware of your own body. Become aware of your pains, sensations, breaths, heartbeat, and so on. You can direct your attention to different parts of your body and feel the life in them.

Your thoughts and emotions. When you are aware enough, you can start to observe the dialogue and observe the drama. You can watch your thoughts and emotions as they flow through you and then freely let them go.

Your connection to the rest of being. At the highest level of pure awareness, you invite the connection you have with the rest of being: the love you have for the waves of the ocean, the admiration you have for butterflies, and the sympathy you have for your fellow humans who suffer around the world. Those connections aren't sensory perceptions of the external world; they aren't feelings of your own body, and they aren't thoughts or emotions. They are pure connections that make you feel you're part of a bigger community that extends way beyond your individual experience of the world.

Take some time now to try it yourself. Explore the furthest corners of your awareness. It is what you were made to *be*.

How to Be

All of us modern-day warriors find it hard to just be. We're trained to constantly do. Our brains wander. Our rushed lives don't give us the time to be present. While there's endless literature and practices that cover the topic, I found those hard to include in my fast-paced life. On one hand, some of the greatest tips came dressed up in mysticism, spoken in dreamy voices interrupted by long gaps of silence, and made up of words that are alien to me. On the other hand, I needed practices that I could take with me on workdays and in airports, not ones that required a quiet place to meditate.

So I developed my own list of practical tips and techniques that helped me and, I hope, can help you reach deep down inside and find your awareness regardless of your daily schedule. These tips will help you *stop* doing by alerting you to your need to become aware, remove the mental clutter, and give you the space you need to observe the world and just be.

Be an Awareness Fanatic

It all starts with making awareness your priority. Be crazy about finding out everything happening around you and inside you. Be curious. Be an explorer. Be a fanatic.

Remember the Selective Awareness Test, when people missed the gorilla walking across the screen as they focused on the basketball? Use your brain's tendency to focus to your advantage. Set out every morning with your brain primed to be open to something new.

As you go through your day tomorrow, try to find out how many different types of trees you come across. For the rest of the week, measure the time your commute takes along different routes. Pay attention to how you treat other people. Notice if you treat your boss at work differently than the people you manage. Monitor your daily water consumption or your posture while you sit. It doesn't matter what you set out to notice, just give yourself a reason to be alert. When you get back home, try to remember as much of your day as you can. If you seem to have forgotten part of the day, spend time trying to remember what happened.

Start a "positive events journal." Stay alert all day looking for the good parts. Write them down. As soon you make them your target, they'll start popping up all over your day, making it a positive, happy day.

When you feel that you're starting to get the hang of it, give yourself the ultimate challenge and monitor the times when you're not aware (because those are events that you should be aware of too). Train yourself to look for the moments when your mind wanders outside the present. You don't have to do anything about it. Just notice and say, "Whoops, my mind slipped for a minute there." The simple act of noticing it will snap you back into the present.

Remember!

➤ **The black belt of presence is to notice when you're not aware.**

Reduce the Distractions

It's hard to stay aware in the modern world because we don't allow ourselves the space. We're often distracted with phones, email, Facebook, and all of today's immersive modern technology. When you're out in public, look around you and count how many people you see staring at the little screens on their devices. Our days are rushed, and we move relentlessly though our endless lists of things to do. When we're given the blessing of a short window of quiet time, we pick up our phones and stare at messages, videos, and posts. When we sit in the car on the ride home we switch on the radio, and when we get home we sit in front of the TV or the Internet until it's time to go to bed. Days pass without a single minute of stillness. Take a stand and reclaim your life.

Remove the distractions. Make it a point to keep your phone in your pocket when you have some quiet time. Switch off the radio on your drive back home and spend time doing absolutely nothing instead of sitting in front of the TV.

Add "me time" appointments to your calendar, short breaks that give you the time to be alone with you. Stick to those appointments. Treat them as you would a job interview. Despite my manically busy life, I found that when I plugged that me time into my calendar before the day filled up and respected it as an important appointment, the rest of my busy schedule fell perfectly around it. I still completed my tasks, but I also kept my sanity with brief moments of presence.

Switch off your data connection, at least on the weekend. When on the web searching, stay focused on what you need, then log out. Dedicate only ten minutes in the morning and ten in the evening for social networks. Get rid of the distractions to free up the space you need to be fully present.

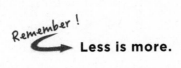

Remember! **Less is more.**

Stop

Yes, that's right. Just stop. Whenever you feel your mind racing or the day rushing by, just stop. Tell yourself that you're not going back to the hubbub of life until you observe ten things around you, one for each one of your fingers. There's a tree, a chubby cat, there's some fresh air, a pain in my left shoulder, and the noise of the water cooler behind me. Count to ten, then take a deep breath and jump back into the day.

Make a Totem

In *Inception*, my favorite movie of all time, the dream world and the real world become entangled. The dreamers use a totem to distinguish between being in a dream and being awake. You can too. Always carry something with you that reminds you that it's time to be aware. It shouldn't be a useful everyday object, but something that's odd enough to serve as a reminder every time you see it. Something simple like a stone with interesting colors, or a spin top, or a yo-yo. Every time you see it, you'll remember that it's time to be still for a short while. When you take your totem out, interact with it. Slow down the pace of your racing brain and be present. I carry prayer beads. When I get them out, I count one observation for every one of the thirty-three beads. I open up and let everything in. A flower, one. The smell of coffee, two. I don't just notice them. I admire them. I relate to them and feel awe for their beauty. I think about how they came to exist and what the story of their life might be. In that state, I don't see a fly as just a fly. I look at the incredible design that lets such a tiny creature perform so flawlessly. I wonder why the grain of wood feels so alive. I think about the probability of random events that might have resulted in them, or the intelligent design that might have interfered. I get fully absorbed in them—and fully free of my thoughts. I become fully aware, resurrected from the dream world of doing back to the reality of awareness.

You can also make a digital totem. Use the home screen of your phone as a reminder. Leave yourself a message there. Set up a few alarms throughout the day with a calming sound to remind you it's time for a bit of presence. Don't let a day go by without those breaks.

Keep your totem in a place where you'll have to bump into it several times a day. I keep my beads in the right pocket of my jeans, and every time I reach into my pocket, I touch them and I remember:

Remember !

➔ **It's time for an awareness break.**

Timeless Time

Give yourself the luxury of a timeless experience at least once a week. I mean "timeless" literally. At the end of the day, take yourself to a quiet spot where you have no access to any time instruments. Go to the ocean or to the woods—or just stay in a quiet room. Make sure you have no connections to time—no clocks, no phones, and no external events keyed to the current time. For the first couple of attempts, this will probably feel alien. Thoughts will rush into your brain and give you a thousand reasons to worry. Hang in there. In time, your brain will give up and give in to the bliss. You will suddenly find everything going quiet and the hours passing by unnoticed. Don't worry about that silly smile on your face. It is a good sign.

Very Important !

➔ **You will be fully present once you remove the connection with time.**

Whatever You Do, Do It Well

Awareness is a state of being, remember. But obviously, we can't just *be* all day. We need to alternate between being and doing to be a productive member of society. Sometimes when what you are doing is too easy for you or when it's too repetitive to capture your attention, you end up zoning out, going through the motions and shifting your attention from the real world deep into your head. There's no reason to lose your presence and awareness when you *do*. You can remain aware by focusing your attention on the process of doing, not the end results.

The trick is in trying to do everything to the best of your ability. Give every little step all you've got and perform as if it's the very first time you've ever done it. Do it better than you did the last time, and take pride that you do it, whatever it is, really well.

This doesn't apply only to your job. It applies to everything, from washing the dishes to spending time with your loved ones. Try it on your commute. Use the time to make your commute an enjoyable and fully aware experience, regardless of the traffic jams. Notice what's around you, or use the time to listen to an audiobook or have a meaningful conversation with a friend. Do something that deserves your time, and you'll wish the commute lasted longer.

Remember!

➡️ **Be aware of the journey. This is where all of life happens.**

Here's one final tip: Do only one thing at a time. Don't watch TV while you eat dinner. Don't spend time with your daughter while "quickly checking your email." Multitasking is a myth. Be fully present.

Remember!

➡️ **Whatever you do, give it your undivided attention.**

The more of these tips you use, or others, to stay aware, the easier it will be to find that peaceful present state. And the more you will wonder how you could ever tolerate the times you spent mind wandering. So take it all in, every experience that life throws at you. Don't miss a thing.

Remember !

→ **Live your life in the here and now, not inside your head.**

Chapter Eleven

The Pendulum Swing

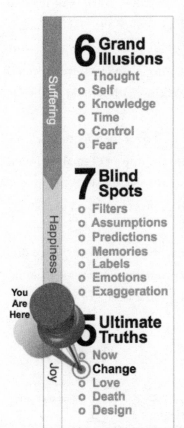

6 Grand Illusions
- o Thought
- o Self
- o Knowledge
- o Time
- o Control
- o Fear

7 Blind Spots
- o Filters
- o Assumptions
- o Predictions
- o Memories
- o Labels
- o Emotions
- o Exaggeration

5 Ultimate Truths
- o Now
- o Change
- o Love
- o Death
- o Design

Suffering

Happiness

Joy

You Are Here

Change is real. The one thing you can accurately predict is that the world tomorrow will be different from the world today. News headlines will capture only the black swans—"Earthquake Hits Island" or "War Kills Thousands"—but they will fail to capture the billions of subtle changes that create those big events, the butterfly effects. From one second to the next, our world changes so drastically that it's safe to say no two instances in the entire history of our universe have ever been identical.

Every subtle change reshapes every instance of our unfolding lives. No change is insignificant. One left turn a second too early might save your life, and a tiny mosquito's decision to turn right can take it away.

A Multiverse

To help you understand how significant every little change is, let's visit a bizarre side of science. Imagine that the little changes caused by butterfly effects could do a lot more than just alter your path. Imagine that each could produce a whole new universe! Over the past couple of decades, scientists have been advocating exactly that in a theory known as the multiverse. You smile to a stranger—a new universe. You frown—a different one. A rock falls—different still. Each of those universes in turn will spawn an endless number of other universes with an endless number of copies of you. In one copy, you're still reading this, while in another, you decide to go buy a coffee and end up finding a different book that changes your path in life and makes you the next president of the United States. Just that tiny change of events could make such a huge difference that the resulting paths would qualify as two different universes altogether.

While the multiverse theory sounds a bit over the top, the idea is a very valuable visual image to demonstrate the impact of small changes. Multiply the far-reaching impact of any small change by the frequency at which such changes occur, and it all becomes too complex to imagine, let alone manage.

Our attempts to exert control over the endless stream of change disappoints us. Regardless how hard we try, our expectations are bound to be missed when any one change initiates a cascade of unexpected, uncontrollable events. And we try harder and harder, to no avail.

We fail to notice how this frantic experience truly looks. We just feel exhausted and

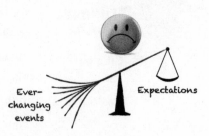

wonder why life seems to be a constant struggle. We don't realize that *we* are the ones who make life harder than it needs to be.

Project Cockpit

Imagine for a minute that advanced technology has enabled us to invent a cockpit that contains all the switches you need to control every aspect of life. Wouldn't it be so cool to keep everything under control? I would be able to control my next promotion, my daughter's behavior, the traffic on my way to work, and every other little detail that affects me every day. But this cockpit would have to be gigantic! With so many parameters affecting our lives, it would have to be the size of a football stadium, with every square inch covered by switches and tiny gauges to tune to keep things under control. At first you think it will be easy to control all the buttons, but once lights start flashing and the sounds start buzzing to indicate all the little changes, you become overwhelmed. When one light turns off, another one turns on. You run faster and try harder, maybe even panic as things keep moving beyond control until you realize that you can't control every switch at once—even if all you need to do is a small turn. Eventually you would fall to the ground exhausted, disappointed by the utter loss of control.

Take it from me, a retired control freak, this was exactly how my life played out every day—as an endless failing quest of disappointment and struggle. Until one day I realized that control is not to be gained at the micro level of every detail. It is not to be found in what I need to do, but rather in how I need to do every little thing I do.

You don't need a cockpit with a million switches. All you need is a couple of simple lifestyle changes. Find the path, and then look down.

Find the Path

Spiritual teachings provide a path that offers a peaceful life. In ancient Chinese teachings, finding a balanced path through the changing faces of life is referred to as the way of the Tao. Buddhists refer to it as the path, and Islam calls it the straight path. Instead of trying to control a million little variables, those teachings recommend that one should simply let most of the events of one's life seek their own equilibrium.

Every factor that affects your life behaves like the swing of a pendulum. As a physical system, a pendulum seeks its point of equilibrium—the point of balance where the pendulum is effortlessly still. You need to exert an effort, apply a force, to take a pendulum out of equilibrium. Once the force is removed, the pendulum will rush back to its natural state, swinging back and forth until it finally settles at the zero point. There, where no effort is needed, the pendulum can peacefully stay in balance forever.

If you want to keep thousands of pendulums steady all the time, let each of them find its own equilibrium. Similarly, if you want to navigate the thousands of little decisions that shape your life, find the equilibrium point for each of them and keep them in balance, away from extremes. You'll need to exert minimum effort to navigate a life of balance. When every pendulum is at its equilibrium point, the line connecting all points is the Path.

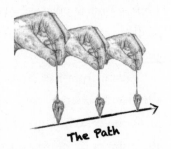

The Path

Remember!
→ **No effort is needed to keep any system at its equilibrium. When everything you do feels effortless, you'll have found your path.**

Extremes exhaust us. Work too much, and you lose the joy of living; work too little, and you suffer from a feeling of worthlessness. Spend too much time with a loved one, and you'll get bored and start arguing; spend too little time, and your relationship will fade. Talk too much, and you'll never listen; talk too little, and you'll never be heard and understood.

Every single thing we do has a point of balance. If you move beyond that point, you'll need to exert effort to keep the system in an unnatural state. Little as it may be, the effort needed to live an unbalanced life adds up exponentially as the number of systems you need to handle increases. It becomes more and more like that cockpit with a million switches—impossible to manage.

Remember!

➤ **Let everything seek its natural balance.**

In Chinese philosophy, the duo *yin* and *yang* describes how apparently opposite forces are actually complementary, interconnected, and interdependent. Everything has both a yin, the feminine or negative principle (characterized by darkness, wetness, cold, passivity, disintegration, etc.), and a yang, the masculine or positive principle (characterized by light, warmth, dryness, activity, etc.). For example, a shadow cannot exist without light, and vice versa. In a harmonious life, yin and yang complement each other. If you throw a stone in a lake, the waves will have troughs and peaks that calm each other down until the water is still again. To find a balanced life, one should embrace both sides and avoid the extremes of either.

Remember!

➤ **Live on the line where the yin meets the yang.**

In Greek philosophy, this approach to balance was described as "the golden mean," the desirable middle between one extreme of excess and another of deficiency. Even the most desirable traits should be balanced. Courage, for example, though a virtue, would manifest as recklessness if taken to excess, and as cowardice when deficient.

Simple as it may sound, this approach to life is almost the exact opposite of what we in the West learn to do beginning in childhood. We learn to take on life, to seek the path of maximum gain, regardless of how challenging it might be. We learn to work to overcome our weaknesses (the hardest path) rather than play to our strengths (the easiest path). We're encouraged to push the boundaries and go to the furthest edges of wealth, beauty, and achievement. As we do, we exert effort, and it makes us suffer.

In my line of work, choosing the hardest path is a national sport. I often meet "power executives" who work long hours, live under tremendous pressure, and fight life in every waking moment. They constantly try to push themselves out of their "comfort zones," and they try to "move the needle." They even try to run their personal lives like machines. Their evenings are reserved for networking dinners and business events. Their kids are driven from tennis practice to music lessons. Every minute is planned and is expected to tick by precisely like a clock. On the rare occasion these power executives allow themselves a short break, they find another extreme, perhaps exercising like Iron Men and Women or running marathons. They push beyond the balance it takes to stay healthy and fit. They may achieve what they set as their target, but they always pay the price.

We often throw ourselves into a fight against life, but in any fight more is lost than won. We're then inclined to complain that life is tough.

Life can be easy. It's the path we choose that's tough.

Remember!
⮕ **Seek the path of least resistance.**

In the movie *Forrest Gump*, Tom Hanks plays the slow-witted Forrest, whose "simplicity" allows him to go through life with minimal resistance. As a result, he ends up being on the All-Star American Football and Ping-Pong teams, meets three U.S. presidents, wins the Congressional Medal of Honor, becomes a shrimp boat captain, creates a large business, and becomes one of the early investors in Apple. Sometimes, like a feather blowing in the wind, the best you can do is travel as the wind takes you. The balance we should be seeking surely lies somewhere in between our hectic modern lives and Forrest's.

Very Important!
⮕ **Live on the Path.**

Look Down

Along with success and progress, one of the core values of our modern culture is ambition. We strive for higher, farther, bigger, and more. We teach our children to measure their worth by how much they achieve not only in the absolute terms of achievement but in competitive, comparative terms. It's not enough to achieve; what matters is to achieve more than another. *That* is what we've come to call success. It's not good enough just to learn; you have to score higher than a peer. It's not good enough to have an enjoyable, rewarding life; you should have a better life than your neighbors do. It's not good enough to enjoy playing football; winning is what matters.

But as we obsessively compare, we set ourselves up for disappointment because there will *always* be someone who's gone farther or done better.

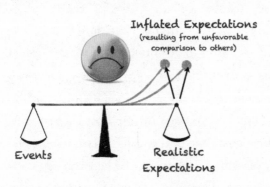

It's not hard to see that we're each dealt a different hand by life. Some are taller and some are shorter, richer and poorer, healthier, funnier, and prettier. That's why if you focus on any specific part of your life, there will always be someone who has "more" than you. We forget the flip side of this distribution curve: each of them has "less" than you in at least one other thing. It's just how the game of life is designed.

Comparing yourself to others who seem to be doing better is a behavior I call "looking up." As we look up, we focus on the parts where we fall short. We try to assess how much farther we need to go in order to catch up with those who lead the pack. We mistakenly think that we're never good enough until we're ahead. As a result, our expectations of ourselves get inflated and consequently missed. Finally, we think that life is unfair to us in comparison to others, and that thought makes us suffer.

There's nothing wrong with wanting to advance in life, but looking up, to compare, will end in vain. There will always be a reason to feel that what you may have achieved is not good enough. Employees look up at managers, and managers look up at chief executives. Models look up at thinner supermodels, and millionaires look up at billionaires.

Here's a challenge: Try reframing ambition so the focus is on the goal of becoming a better person regardless of how you compare to others. Even better, *Look down.* Work hard, grow, and make a difference in the world, but please feel good about yourself. Please stop looking at what you don't have. What you don't have is infinite. Making that your reference point is a sure recipe for disappointment—and a sure way to fail the Happiness Equation. Instead of looking at the few who appear to have more than you, look instead at the billions who have less. Yes—billions!

If you can afford to buy a coffee for a couple of dollars, be grateful, because more than three billion people live on less than $2.50 a day, and more than 1.3 billion people live on less the $1.25 a day. If you can drink a glass of water, be grateful, because 783 million people do not have access to clean water. If you have a home, be grateful, because there are close to 750,000 homeless people freezing on the streets of big cities just in the United States.

And it's not just material wealth. If you look closely, you'll find that sorrow and misfortune—though hidden—are a lot more pervasive than you thought. Perhaps the most beautiful example of how we miss other people's sorrow is captured in the mystery of the Japanese smile. While a smile for most people expresses happiness, for the Japanese a smile can express a variety of feelings, including awkwardness, doubt, fear, shame, and embarrassment. In Japan's silent culture, it's not customary to express extreme emotions, especially negative ones. If a person makes a mistake, for example, he or she smiles. This smile is used to mask the feeling of shame. I once asked a friend why everyone in Tokyo constantly smiled when I knew that the infamous pace of life there caused a lot of hardship. In her beautiful words she said, "We keep our suffering for ourselves and give our smile to you." I truly admire Japan, and it intrigues me that an entire nation can—with so much dignity—hide its pain.

There's so much sorrow to go around, so if you have to compare your life, flip your outlook upside down and compare yourself to those who are less fortunate. When you change your perspective, you'll see many reasons to be happy with your blessings.

A friend of mine, a successful and ambitious businessman, was always shooting for higher goals. Then he was diagnosed with acute pancreatitis, a disease that causes the stomach acids responsible for digesting food to spill inside the cavity of the abdomen and digest the patient's own flesh instead. For months, he lay on a bed with tubes puncturing his body, kept alive by drugs and drip liquids. As his health deteriorated, his ambitions dwindled. He was no longer interested in material gains or career growth. He stopped comparing himself to the person who got promoted ahead of him or the neighbor who drove a fancier car. When he finally stabilized, his ambitions shifted from acquiring the next material success to, in his words, "being able to turn on his side in bed."

Very Important !

Only when we *look down* do we realize how fortunate we really are!

Looking down helps you appreciate the good things in your life. And it's not a secret that the feeling of gratitude makes us happy.

Psychologists Robert A. Emmons of the University of California and Michael E. McCullough of the University of Miami conducted a study in which they asked three groups of participants to write a few sentences each week focusing on a specific topic. One group wrote about things they were grateful for; the second wrote about things that displeased them; and the third wrote about events that had affected them either positively or negatively. After ten weeks, those who wrote about gratitude felt better about their lives. They also exercised more and visited physicians less than those who focused on sources of irritation.[1]

Martin E. P. Seligman of the University of Pennsylvania tested the impact of gratitude with hundreds of participants. Each was initially asked to write about an early memory, then to write a letter of gratitude every week and personally deliver it to someone they wanted to thank. Participants exhibited a huge increase in happiness scores when they expressed gratitude, and the impact sometimes lasted as long as a full month.[2]

Remember !

➤ **Gratitude is a sure path to happiness.**

And *grateful* is a matter of mind-set. When you look down, you learn to invite more gratitude into your life. You may even learn to be grateful for your own sorrows when you see that there's always someone with deeper wounds. In comparison, you realize that, by a stroke of good luck, you have been spared.

Take my own example of losing Ali. Can I look down and feel grateful even with such a tragic loss? Is there anything worse than losing someone as wonderful as Ali? Of course! It could have been much worse. Many young men in their twenties are diagnosed with cancer, for example. They endure the harshness of chemo and radiation therapy for months, and still some don't make it. Would that have been a better way for Ali to go? No! Some university students mix with the wrong crowd, end up addicted to drugs, and die of an overdose. Would that have been a better way for Ali to go? No! Even simpler, would we have preferred if Ali's fate was exactly the same but back in Boston, where he lived and studied, instead of at home, where he came to visit us for a few wonderful days before he left? Of course not!

If I want to *look up* at how much longer we could have lived to-gether, I would suffer because the fact is he left and there's nothing I can do about that. Instead, I choose to *look down* and feel grateful for the twenty-one wonderful years during which he blessed us with his presence. **Instead of feeling resentful that he died, I feel grateful that he lived.**

"Ali was never afraid of much," a friend of his once told me. "He was uncomfortable with heights, but not much truly scared him. I re-member asking him what his greatest fear was, to which he responded that his greatest fear was losing someone he truly loved. This included his family and his closest friends. When he left, I realized what seemed like the inevitable never happened to Ali. He got to live without his greatest fear ever coming true. Which I think is pretty spectacular."

When it comes to leaving this earth, Ali's dying peacefully in his sleep among his loving family was not the worst-case scenario. If you look down, there are countless scenarios worse than the one we lived through. Like everything in life, while having him stay one day longer would have made life better, I understand that even this could have been much worse!

It's now your turn to reflect on your own hardships. As you do so, be fair and realize that while you might not be the luckiest person alive, you surely aren't the unluckiest. If you ever forget, then please:

Remember!

Look down!

Chapter Twelve

Love Is All You Need

6 Grand Illusions
- Thought
- Self
- Knowledge
- Time
- Control
- Fear

7 Blind Spots
- Filters
- Assumptions
- Predictions
- Memories
- Labels
- Emotions
- Exaggeration

You Are Here

5 Ultimate Truths
- Now
- Change
- Love
- Death
- Design

Suffering · Happiness · Joy

I love butterflies. I don't care what kind, what color, or what size. I just love them. I don't want to own them, and I don't even necessarily have to see them. I'm just happy they exist. I love them so much that I want to hug them. But I don't. I just pour that overwhelming feeling of love over them whenever our paths cross. I guess they know because they do seem to come my way a lot. Sometimes as I walk to work, a butterfly will start to fly gracefully right in front of me. It will land calmly on a branch in my path, as if to say, "I'll stay right here and pretend I'm not looking so you can enjoy me."

As I pass, it flies around me

and then lands again in front of me. I would not stop, and neither would she. Coincidence? I think not. Either way, I don't care because I just *love* every single butterfly that has ever existed.

I also *like* them. I like their patterns, beauty, and grace. I admire their life journey from caterpillar to beauty queen, their passage through the hardship and uncertainty of the cocoon. I appreciate the work they do for us pollinating flowers, and I respect them for the perseverance they show in their short lives despite their fragility.

Liking, admiring, appreciating, and respecting are all different feelings, and they are all different from love. I like and I admire for particular reasons. Love, on the other hand, is just there: unexplained, unsupported by any reason, and unchanging.

Love—true love—is real. All other emotions are temporary. They appear when a reason triggers them, and they disappear when that reason goes away. Liking the patterns of a butterfly depends on how beautiful a particular butterfly is. If it happens to be a pale gray, unattractive one, the admiration might fade. My *love* for it, however, remains.

Think of the nearly universal experience of motherly love. After suffering the uncomfortable months of pregnancy and then the acute pain of labor, a flood of love overwhelms almost all new mothers when that tiny crumpled thing is placed in their arms. This overwhelming emotion endures regardless of circumstance. A child may grow, leave home, and never call—but the love remains just the same. The child might even leave this world altogether, as Ali left us, but the unwavering motherly love only grows.

Which Type of Love?

Pop culture makes love seem like a painful endurance test. Heartbreak and longing inspire countless lyrics and fill the pages of novels. Yes, there is a type of love that causes pain. The other type, however, leads to pure uninterrupted happiness. The type of love in pop culture is an illusion, while the type that is more rarely spoken of—but so much more deeply felt—is real.

Conditional love is driven by the thought "I love *because* . . . ," and like everything that is based on a thought, it's an illusion, it's impermanent, and, as the thought evolves, it will eventually and inevitably lead to suffering.

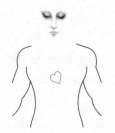

In contrast, *unconditional* love is felt but not understood. It's genuinely built upon "I love" and nothing more— no reasons or preconditions, no expectations and no demands, and consequently no disappointments. No thoughts! This is the only form of true love. It's rare to find, but it's real.

Remember ! ➤ **Unconditional love is real. It's the only emotion that's not generated by a thought in your head.**

A Feeling without a Thought

The true essence of what makes unconditional love real is that it lives outside the realm of thought. All other emotions are based on a thought.

Envy is generated by the thought *She has what I don't have.* Hate originates from the thought *Something about that person strongly contradicts my life model.* Admiration stems from the thought *I have analyzed a person's attributes and find that they exceed my expectations.* Anger is sparked by the thought *The behavior of another threatens me. I need to be forceful to feel safe.* The thought triggers the emotion.

Conditional love too originates in a thought: *She makes me feel happy, and therefore I love her* or *He makes me feel safe, and therefore I love him.* This same model applies to our love for things: *This car makes me look good, therefore I love it* or *My shoes are comfortable, therefore I love them.* As long as the reason persists, the conditional love lingers, but once the reason goes, the thought pattern changes, and the feeling fades away. It may even morph through incessant thought into hate, anger, fear, or any other thought-driven emotion.

This is the reason relationships suffer: they're built on conditional love in an ever-changing world. Expectations of beauty, entertainment value, physical pleasure, and other forms of expectation have become preconditions for love. When the lover changes, the expectations are missed and the fairy tale turns into a nightmare.

Unconditional love, on the other hand, withstands every change. It cuts through the Illusion of Time. Even when I don't see my wonderful daughter for months, an unchanged, ever-growing love for her always fills my heart. True love cuts through the Illusion of Knowledge. We love the ocean, the stars, the birds, and the beasts; we feel a connection to them despite not understanding their ever-changing nature. It cuts through the Illusion of Self by allowing you to love that which is too far away for your physical senses to grasp. It's the only form of love that is everlasting, extending even beyond life itself. Ali's love will always remain in my heart even though he moved on and left our physical world.

I use the word *always* here intentionally. True love is *always* felt, every second of every day. Time is not a condition for a love that needs no conditions.

The True Joy of Love

There's no happiness without love. And while conditional love often causes suffering, true love delivers lasting joy. There is no taking in true love. With nothing to take, there's nothing to expect and none of the suffering that results from the missed expectations we all know from conditional love. Wanting a reward, gratification, and even just to be loved back are all conditional. *I deserve to be loved as a precondition for my happiness* is an ego-driven thought, just another attempt to prove that we're "good enough" and so worthy of being loved. Any thought that stems from ego is bound to disappoint, and as that illusory love fades, suffering will set in. But with no expectations—no demands from your beloved—the joy of love settles in because:

You Feel Happy Because Your Love Comes with No Expectations

Remember!

⮕ **"No expectation" never turns into a missed expectation.**

Of all there is to know about unconditional love, its impact on your happiness is surprisingly simple:

Very Important !

→ **The true joy of true love is in giving it.**

The Economics of Love

Fill the world around you with love and it'll bulge out of shape, shake with instability, and give you more of itself than you expect. Just try it, then see what happens.

Very Important !

→ **The more love you give,
the more you get back.**

I wish there were scientific studies to prove this so I could dazzle you with a chart or an impressive stat. Love is not a widely researched topic in science, but consider this analogy: In physics, the law of conservation of energy means that energy never goes away. It never even diminishes. It changes form, but in any closed system the amount of energy you start with will be the amount you end with. Love follows the same law: true love can't be destroyed; it just changes from one form to another. Because of that conservation, the love that you inject into a system morphs and then comes back to you from where you least expect it. Actually, it does one better than energy: it attracts the love of all beings to you. Like a savings account, the more love you deposit, the more it grows and multiplies so that when it's time for you to withdraw, even more will be there for you.

Call that the law of conservation—or multiplication—of love.

Very Important !

→ **Love never goes to waste. The more you give it away, the more loved you will feel.**

Look at those individuals who have peacefully and unconditionally loved the world and everyone in it even as it brought them suffering: Mother Teresa, Gandhi, His Holiness the Dalai Lama. Billions from all faiths, lands, and walks of life love them. That love for them continues long after they pass away. We love them even without knowing the details of their stories. Even you must love someone like them. How could you not?

Never in business have I heard of a free and renewable resource that offers such spectacular returns as love. It's a bit like the economics of rock music stardom. A talented musician can sit alone in a room and create a masterpiece from nothing but his own inspiration, and earn admiration and fortune for decades to come. While that kind of talent might be very rare indeed, we are all capable of creating masterpieces of unconditional love. Such a powerful commodity requires some special handling. I follow three practical tips I call Love's Instruction Manual.

Love's Instruction Manual

Here are my secrets for how to benefit from the spectacular economics of love.

Love Everything and Everyone

A snake may look creepy and behave in a sneaky way, but it's not evil; it just meticulously does what it was made to do. It never does more of it than needed, and it never misses a beat. If you hate snakes, what you hate truly is just the story you made up about them in your brain, the story that says they're evil and slimy. But they're not. No snake out there attempts to hurt anything for the fun of it. It hunts to eat, just as we do. While most of our own hunting now is in a supermarket aisle, we're no better than a snake when it comes to survival. We're the biggest carnivores of all species by far. Yet we all certainly feel we're worthy of love.

Take away the thoughts, the preconditions of how you would prefer a snake to look or behave, and what are you left with? No reasons for material emotions, just unconditional love. Avoid snakes to stay away from harm, but don't hate them just for being themselves.

If you can love a snake, then you can love all other beings—the trees, the rocks, and the bees. Even when it comes to human snakes, if you look behind the mask of ego, nothing remains but pure love. Even the most annoying, seemingly hateful people you'll ever meet, when you see behind their egos, fears, and thought-obsessed behaviors, you'll find peaceful children who just want to be loved and appreciated. Once loved, most of them drop the masks and turn real.

Remember!

➤ **Gently remove the mask of ego and love what you see underneath.**

Idealistic as this may seem, I also am realistic. I know that humanity has given us examples—tyrants, murderers, and evil villains of all sorts—that make the idea of unconditional love hard to believe, but these are exceptions not the rule. I've worked with some of the world's most difficult politicians, and even in them I found many who are human inside.

When it comes to those (very few) who are so deeply locked in ego that their true selves never surface, I learned a very effective strategy from Ali when he was a young child. He would give people with an unshakable ego three chances. After that he would avoid them or tell them openly, but politely, that they were simply not compatible. But even when he stopped dealing with them, he would still love them, and, I'm certain, somewhere deep inside they would love him back.

Please note that loving everything and everyone is not a naïve, romantic, or idealistic approach to life. As a matter of fact, this strategy is a bit selfish. On top of all the love you get back, unconditional love solves your own Happiness Equation. It gives you the joy of love, which is found in giving love and expecting nothing in return. No missed expectations. Just peace. It's a wise choice indeed!

Love Yourself

How can you love anything, or expect anything to love you, if you don't love yourself?

Nothing causes more unhappiness in the Western world today than the widespread deprivation of self-love. Studies show that only 4 percent of all women in Western societies believe they are beautiful, and more than 60 percent believe they need to be thinner to deserve to be loved! Sadly, this shouldn't be surprising. We are systematically trained not to love ourselves unless we meet stringent expectations.

As a success-obsessed society we've grown to believe that being average—being like most other people—is not "good enough." When you think about it, this speaks to a tremendous arrogance in its suggestion that most people are not good enough! An average figure is not attractive enough; we need to be supermodels. But even supermodels don't feel good enough because there's always a more attractive supermodel. Being average feels threatening because it means those who are above average will deprive us of the opportunity to succeed in a competitive world. But it goes without saying that we can't all be above average. This would contradict third grade math. Someone needs to be above and someone below for there to be an average in the first place!

Pushing oneself unrealistically is a sure way to rack up missed expectations, disappointment, and suffering. In other words, it's a sure way to mess up the Happiness Equation. With the mounting disappointment, the stress compounds itself until we can no longer take it.

Lack of Self-love Makes Us Feel That We Always Miss Our Own Expectations

Please stop here for a minute and ask yourself if this is the way you treat those you love. No, you give them warmth and reassurance. Then why would you treat yourself this way?

After all, you're a mammal. And mammals instinctively care for newborns, who are not yet ready to take on the world. This programs us to seek out and desire those feelings that keep us safe when we're vulnerable. The warmth, soft touch, and gentle communication we get from our parents as newborns reduce our stress. When we feel safe, our brains trigger the production of feel-good hormones that allow us to perform better and be happier. This is the way we should care for ourselves too. Treat yourself as you would treat a loved child. Give yourself warmth, love, and tenderness. No good can ever come from harshness. Love is all we need.

Self-love works, and it's attainable. And that too I learned from Ali.

The thing he did best was accepting himself for who he really was. He would always try his absolute best—and then appreciate himself for trying regardless of the outcome. As long as it was the best he could do, he would never blame himself for failing to hit a particular target. He excelled in music but not in sports, and that never bothered him. He was lucky in friendships but not always lucky in love, and that was fine because it was who he was and he liked that. Everyone who met him liked it too.

Remember!

➤ **Love yourself for doing your best.**

But that's easier said than done because we tend to remember what we don't like about ourselves and how others criticize us more than what we *do* like. This bias sways our appreciation of ourselves the wrong way, but there's an easy way to fix it. Start a journal or simply send yourself emails. Either way, jot down everything about you that's positive and admirable. Force yourself to write at least one thing a day that you're proud of. Write down every compliment you receive: what it was, who said it, and what prompted it. Go back and visit your journal whenever you feel that you're not good enough. It will expose the negative thoughts and remind you that you're really not that bad at all.

More important, surround yourself with people who make you feel good about yourself. Never let bullies or destructive critics into your life, not even for a minute. Be open to positive, constructive feedback that's given with love, compassion, and care, but shut out the flatly negative. If a friend shows any sign of that negativity, do as Ali taught me and give them three chances. Tell them:

Very Important !

➤ **What you just said made me feel bad about myself, and I don't like being around people who make me feel bad, so, don't do it again.**

If they continue to be negative, draw that boundary again clearly, and if they do it a third time, pack up and leave! Tell them directly:

Very Important !

➤ **You make me feel bad about myself. I deserve better!**

Even if they beg and plead, don't turn back. Three chances are more than enough. Assertiveness will save your life and will also help teach them to treat their other friends better.

Finally, remember that no reasons are necessary to love yourself *unconditionally*. You're not your ego. You're not your achievements or possessions. You're not the success or status or anything that you demand from yourself as a precondition for self-love. The real you, Pooki, deserves to always be loved.

Remember !

➤ **You—apart from your ego—are truly lovable.**

Be Kind

What do you do when you truly love? You willingly give. Giving something to the one you love feels as good as keeping it for yourself. It often feels even better.

If you learn to love everything and everyone, then give unconditionally. Give a few cents to a charity or drop a dollar in the hat of a performing street artist. A single dollar can feed a family for a day in the developing world, so skip your daily coffee once and feed a child for a week.

But do give more than just material gifts. I invite you to give a smile, a word of appreciation, a good conversation, or a compliment. Give love, acceptance, and understanding with no judgment. Acknowledge those who cross your path: a waitress, a shop assistant. Don't treat them like two-dimensional beings, objects there to serve you. Respect your elders. Introduce a friend who needs a contact. Pass on a CV to your HR Department. Call those going through a tough time and just listen. Help if you can. Make them feel that someone cares. **Treat everyone as you would like to be treated.** That's the golden rule of love.

Extend your gifts beyond just the people around you. Give water to a tree, pet a cat, feed a bird, spare the life of a fly. Care for your car, your books, even your coffee mug.

If you give, life always gives back to you. Think of it as if you were giving to the whole universe when you give. It owes you and will pay you back with interest! Nothing goes to waste.

Remember! **To love is to give all you can.**

Give what you don't use. Shoes, jeans, and dresses are made to be worn; if you're keeping them in a closet, you're killing them. Give them life by giving them to someone who will love them and use them.

Life thrives when it flows. A life of giving is like a river, fresh, flowing, and full of life, beautiful and happy. Water kept motionless is just a swamp, stale and sad. Which would you like to be?

Remember !
⮕ Let life flow. Keep what you use and give the rest away.

When you do give away something you love and value, the universe pays you back with interest. While that may be hard to grasp at an individual level, it becomes clearer when you consider society at large. In economics, we know that if those who enjoy abundance gave to everyone in need, the whole economy would grow and the givers would benefit and reap more than they gave as a result. This is why, during downturns, economists and policymakers urge consumers to keep spending. It sounds counterintuitive to consumers' instinctive tendency to save during tough times, but here's the catch: if they stop spending, the world comes to a grinding halt, but if they continue to spend, society prospers in the long term. The more we give, the more abundance we create.

Very Important !
⮕ Giving never takes away from the sack. More always comes back.

Furthermore, the wider the circle in which you spread your gifts, the more your returns will multiply. When you help someone you don't even know, you do so with no expectation of payback—and that's when you hit gold. When we give selflessly, life itself assumes the debt and pays back generously with its unlimited resources. Take it a step further and give someone you don't even like the gift of a kind word or the gift of withholding judgment, and the virtuous cycle continues.

Over the years I chose to believe this to be true. My engineer's brain got curious, and so I decided to test the system. Whenever I gave, I took note and found myself receiving gifts from unexpected sources, often gifts with value that far exceeded what I gave. When I stopped giving, life became harder and it felt as if I needed to struggle for everything I earned.

In a study conducted at Harvard Business School, Michael Norton, Elizabeth Dunn, and Lara Aknin gave strangers a sum of money ($5 or $20), and told them to spend it the same day. Half were told to spend the money on themselves, and half were told to spend it on others. Those who spent the money on themselves bought things like coffee and food, while those who gave the money to others reported spending it on things like gifts for their siblings and donations to the homeless. The result? Those who had spent their money on others reported feeling much happier at the end of the day than those who had spent it on themselves. This was consistently the case regardless of how much money was spent.[1]

Several similar studies confirm that **money can indeed buy you happiness—if you give it away**. So does everything you have to give: your smile, your time, attention, knowledge, compliment. In that sense:

Remember!

→ **Giving is the good side of selfish! It makes the giver happy.**

Last but not least, the ultimate form of giving is *forgiving* those whose behavior doesn't seem to warrant it. Forgive the driver who cuts you off in the morning, the coworker who backstabbed you, the "friend" who posted a nasty comment on your Facebook timeline.

There may be a thousand different and perfectly good reasons why the driver behaved the way he did. Maybe his wife is in labor and he was rushing to be with her; maybe a horrible driving school instructor trained him; maybe he was avoiding another car that cut him off; or maybe he was attempting to save the life of a squirrel crossing the street. At some point, you will be in his shoes. Forgive and you will be forgiven. It always comes around.

Forgive those who argue, even when you believe them to be wrong. One thing I loved most about Ali was how he yielded in arguments even though he was not convinced. He would listen attentively and then make his point. He had no urge to prove that he was right, but he did have an uncontrollable urge to be kind. I assure you, the whole world was kind to him in return.

Very Important !
→ Choose to be kind instead of right!

Well, this was my sentimental, unscientific chapter. But perhaps it is the most appropriate chapter for you to forgive me for that. Unexplained (and unexplainable) as it is, unconditional love is one of the pillars of our universe. When it comes to finding your state of joy, the Beatles said it best:

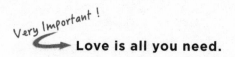

Very Important !
→ Love is all you need.

Chapter Thirteen

L.I.P.

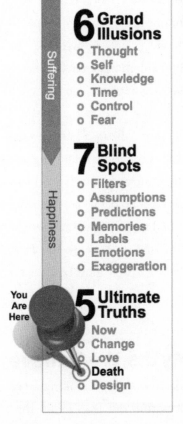

6 Grand Illusions
- o Thought
- o Self
- o Knowledge
- o Time
- o Control
- o Fear

7 Blind Spots
- o Filters
- o Assumptions
- o Predictions
- o Memories
- o Labels
- o Emotions
- o Exaggeration

5 Ultimate Truths
- o Now
- o Change
- o Love
- o Death
- o Design

Suffering

Happiness

You Are Here

Death is real. No one who has ever lived has escaped death. It may even be more real than life itself.

This chapter has been the toughest thing I've ever had to write. There is an eeriness surrounding death—and, as you can imagine, this is particularly difficult for me now.

Death scares us so we don't talk about it. But not today. In this chapter I am going to be brutally honest in addressing the subject. Please accept my apology in advance if parts of it sound harsh or conflict with your belief system.

The finality of Ali's death has brought me face to face with the fundamental truth of life and death. As it crystallized my understanding, it intensified my commitment to live a life worth living. It also took away my last fear: I'm no longer afraid to die.

Perhaps as we explore death, you'll find, as I did, that our biggest fear isn't warranted after all and that within death resides our most valuable life coach. This won't be an easy read, but it'll sure be worth your time.

Most of us in Western-influenced cultures avoid discussing death. As a result, there are many things that we don't know about it, and that, in turn, scares us more. Many other cultures, however, speak openly about death, and some even have traditional celebrations using its imagery. Mexicans, for example, celebrate their deceased during the Day of the Dead festival once a year. These celebrations are very different from most Western traditions in one crucial way: this special occasion is an invitation to party *with* the dead, not *for* the dead. The dead are present, not simply remembered or commemorated. There's an abundance of food, presents, and flower decorations. The living tell stories about their loved ones and welcome the souls of the departed as they come to visit. In a similar fashion, Sufis throw a party on the anniversary of someone's death that includes traditional whirling and feasting. In Rajasthan, after twelve days of mourning, survivors throw a party for their departed loved ones. The Irish hold a wake, a raucous celebration of the departed with laughter and music. How can there be such a diverse set of views around the same topic? Surely people from those cultures don't die differently than we do. So it must be a matter of perspective, a different point of view.

The Myths We Believe

If we allow ourselves to look closely at death instead of looking away, we might find a meaningful place for it within our lives rather than view it as an enemy at the gates. The first phase of this process is to dispel a few myths.

There's a Day to Die

Dying is an integral part of the process of living. The minute we're born is the minute we start to die. You are dying as we speak. Your red blood cells, all twenty-five trillion of them, will die within the next four months. By the time you finish this chapter, more than a 150 million cells all over your body will die. Of those, two thousand will be brain cells, which will never be replaced. Death is not an event—it's a process. There's nothing special about the day we go.

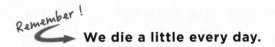

Remember! **We die a little every day.**

Death Is the Enemy

Death is an indispensable part of the food chain that sustains every life-form on this planet. Each species preys on the death of a subordinate in the chain. Without the death of another in the system, life wouldn't survive. We humans prey on most other creatures until our own death, when a plot of grass and, perhaps, a rose tree will find nourishment in our decay.

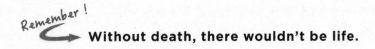

Remember! **Without death, there wouldn't be life.**

Some life-forms stick around longer than others, but all, without exception, eventually go. Every minute, billions die peacefully after playing their roles in keeping the ecosystem alive. The only species that makes a big deal of it is our own.

Death Is Always Unwelcome

Deep inside, we all know there's no escape. We do wish, however, that death would book an appointment instead of showing up uninvited. And we would prefer a late appointment. "Hey, I'm still healthy and enjoying life here, so come back in thirty—no, 330 years. Actually, don't bother. Leave your number and I'll call you when it's time."

When life is going well, we never want to die. Many of us, however, have experienced the other side of that. When life goes against our wishes, our attitudes can change. In cases of a painful illness or when our bodies becomes old and frail, we start to wonder, even if we find it hard to admit, "What's taking death so long?"

We disagree with death on timing, and that applies to the mortality of those we love as well. When a loved one dies, we feel betrayed. We get angry at death for taking our loved ones too soon. "If only we could have one last hug," we think. Ask yourself this question, though: How soon is too soon? What if I had managed to strike a bargain that had kept Ali for one last hug, or even another year? Would I have then said, "Okay, you can take him now"? No! It still would have been too soon. I would have always longed for one more hug.

But when life is no longer the better option, our disagreement with death fades away. When the extensive bleeding that Ali suffered started to affect his vital organs, they failed one by one. For hours I kept hoping, pleading, praying that he would recover. I believed that it was not yet his time, but when an honest doctor finally took charge of the ICU and informed me of the extent of the damage Ali's body had sustained, my heart shifted. I wondered if staying longer in our world to suffer with a damaged brain was a better route for my handsome son. Perhaps that moment was a better time to leave than later.

Like it or not, death has already booked an appointment with each of us. It just hasn't told us when it is. Perhaps this is what allows us to enjoy our time being alive. When death finally does show up, however, our minds may have shifted. We weigh the alternatives and may feel that we're ready.

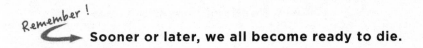

Remember ! Sooner or later, we all become ready to die.

Death Is Painful

Another disagreement we have with death is the question of how we will die. We think, I don't like drowning; it's, well, too wet. I don't like falling either. Perhaps there's a way to die from candy? Yes, that seems more like it. Cotton candy—that is how I want to go.

We get angry at the world, at God if you will, when a tsunami takes the lives of thousands of people. It feels cruel. Surely there's a better way to die. But when it comes to death, it's always sudden and always resented. It makes no difference how.

Ali always told me that he wasn't afraid to die, but he feared the pain of dying. I remember he brought the topic up when he was eleven. (I guess he had to bring it up early because he had to squeeze a full life into just twenty-one years.) My answer to him then was "Make it your wish, *ya habibi*"—my loved one—"that you never have to suffer that pain." On the day of his departure he went to sleep at 10:30 p.m. and has not yet woken up. When the time comes, my wish, like Ali's, will be to go the same way, peacefully in my sleep. That is even better than cotton candy.

A painful death is one of our biggest fears, but should it be? There is no painful death, only a painful life in its last moments before death. Think about that. When we go, there is no more pain. As Woody Allen said, "I'm not afraid to die. I just don't want to be there when it happens." He won't. When it's our time to die, none of us will be there.

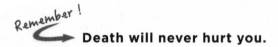

Remember ! **Death will never hurt you.**

Death Can Be Cheated

This myth is the invention of our modern world. Before the promise of "saving lives" became the foundation of the multitrillion-dollar health care industry, dying used to be a lot simpler. Now it takes a lot longer. It's more complex, more painful, and vastly more expensive.

In the past most people died suddenly or over a short period of time. Death was expected and accepted, and while it still left loved ones behind in shock and grief, it was easier for the dying. That's because there was less prolonged pain, less suffering. Now times have changed. Constant advancements in technology are dedicated to finding a cure so that when something goes wrong we can fix it. Billions have been given a chance to live another day, and as a result global life expectancy shot up by 50 percent in the past sixty years alone.[1]

Sometimes, however, living longer does not mean living better. Peter Saul, an intensive care specialist, discusses this dichotomy in his TED Talk, "Let's Talk about Dying." The promise of "saving" lives is inspiring, he says, but the product on offer should more accurately be called "prolonging" life. When thought of that way, living another day becomes a precious gift only when the life is worth extending. But many in the advanced world don't make this distinction and bargain for more years, even when that means more suffering. As a result, one in every ten people dies in intensive care, and patients are kept hooked to life-support equipment even after they're declared clinically dead.[2]

As we live longer, another multibillion-dollar industry is tasked with injecting sticky liquids in us and cutting, stretching, and stitching our skin to remove signs of aging. Then, when we finally die, yet another industry offers us immortality by freezing our body in the hope we can come back to life when technology is advanced enough, a promise as ancient as my fellow countrymen, the pharaohs.

If anything, it's a miracle we're alive in the first place. Take a look at your physical form and think of the hundreds of vital functions that need to operate flawlessly to sustain its continuation, just from one moment to the next. Think of the tens of thousands of proteins contained within it, each performing as a sophisticated device. Think of the trillions of cells that need to be fed, protected, and replaced. We wake up every morning expecting this machine to function as a simple routine, but the truth is, our bodies are extremely fragile. If a single genetic pair morphs, a single germ persists, or a single vital organ fails, if any single system breaks, the machine will collapse. There are countless points of possible—probable—failure.

Think of the handsome form we called Ali. His death was the result of a pinprick! This truly is what it was: a needle puncturing a blood vessel. Are we really that fragile? We are! Much more than we care to think. So many things can go wrong and often do. As the Arabic proverb puts it, "The reasons are infinite, but death is one and the same."

With Ali's departure, I've made amends with death. I know for certain that this could very well be my last breath; it could be the last paragraph I write. This machine, my physical form, comes with no warranty. Its instruction manual clearly states that I have no say about when it will cease to function. With meticulous maintenance, we may get a few thousand miles more on the clock, but eventually we run out of spare parts. It's just the way the game is played.

Remember !

➤ **There's no cheating death. We all go someday!**

When something resides so firmly outside our scope of control, it becomes impossible to find happiness until we learn to **accept it** as an integral part of the normal course of life. And yet it's hard to accept death when we believe that it takes away life itself. But does it?

The Long Life Continuum

When Nibal and I were finally allowed into the intensive care unit to say our final good-byes to our son, she told him, "*Habibi*, you are finally home." I kissed him on his forehead and said, "I will see you again soon, my friend." We were already at peace. Our state resulted from a conviction that some readers will not agree with: our belief in the afterlife.

Definitions

There are different schools of thought on what happens to us after death, but there are some recurring conceptual building blocks. The most common are those of eternal life, reincarnation, and nothingness. Religious belief systems generally say that we live on in eternity and end up in heaven or hell—a view that assumes that true life starts only after death. Other belief systems advocate a less dualistic approach and say that we return to live again and again. And the secular belief system says that there's "being" and "nothingness," and when we die it's the end: we dissolve away.

None of those diverse views can be confirmed with any degree of certainty. But to find a common ground, let me suggest a unified definition that cuts across all of them for what *life* is. I use the word *life* here to refer to life in our current physical form on this planet and the word *death* to refer to the end of this form. No controversy there. With these two definitions locked away, we can turn to something new: *long life*. This is how I refer to the combined duration of life and whatever your definition is for what happens *after* death. This equals (Life) + (Eternal Life) if you are religious, (Life)*(Cycles of Reincarnation) if you believe in nonduality, or just (Life) if you are secular.

A Question of Time

Long life depends on rethinking some things we take for granted. For instance, we understand death as an interruption of life, but it seems

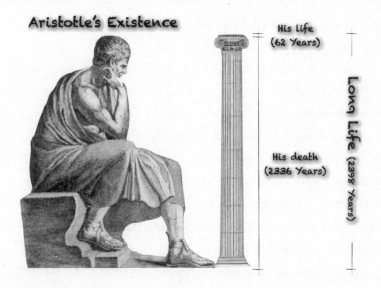

Aristotle's Existence

His Life
(62 Years)

His death
(2336 Years)

Long Life (2398 Years)

obvious from a different perspective that, in fact, it is life that interrupts death. Death lasts infinitely longer, while life ends so quickly. How can something as precious as life be so insignificant? How can it last for such a short time? And why do we place so much weight on that one little piece of the infinitely large picture?

Well, the answer resides within the relationship between life and time. Luckily, we do not have to rely on metaphysics here. Regular old physics has been looking at the large, the small, and everything in between for the past century and a half. The results—quantum theory, the big bang theory, and the theory of relativity—can help us understand why we look at life and death the way we do. This is good news because life after—and before—death has been the center of debate since the dawn of humanity, so the ability to discuss this topic with a level of objectivity is welcome indeed.

You've probably heard of the double-slit experiment in quantum physics. For our purposes, I'll just say that it represents the only link I know between physics and the nature of life itself. It associates the existence of subatomic particles to observation by a life-form (such as yourself). In this simple experiment, subatomic particles—photons, for example—are shot at a wall through a barrier that has two slits side by side. A single particle, in the absence of observation, will pass through both slits, at the same time ceasing to exist in its particle form and becoming a probability wave function. Only when observed does the wave function collapse and revert to its physical form as a particle, which then passes through only one slit or the other. Simply watching the photon seems to make it "choose" to be an actual particle. This bizarre characteristic has been the topic of ample studies, all of which point to one confirmed conclusion: **When unobserved by life, the physical world ceases to exist!**

Schrödinger's cat is a famous illustration of this. In this thought experiment a cat is placed into a steel chamber, along with a device that contains a substance that might kill it. The device is activated to release the possible poison via a random event that we can't control and cannot predict from outside the box. Since we can't know, according to quantum laws, the cat can be in any state on the probability wave function and accordingly is *both dead and alive* in what is called a super-position of states. It's only when we break open the box and observe or measure the condition of the cat that the superposition is lost and the cat becomes one or the other, alive or dead. This is the Observer's Paradox: the observation creates the outcome, and the outcome doesn't exist unless the measurement is made.

Heisenberg's uncertainty principle pushes this strangeness further and proves that our own act of observation changes the reality of the world we observe. The uncertainty principle suggests that the physical world—the world all around us—is observer-dependent. Without an observer, in other words, everything would remain a wave of endless probabilities. You and I and every other *life*-form are not a product of the physical world; it is a product of us, because by observing it, we make the physical world what it is.

Yeah, I know, this still freaks me out every time I ponder it.

Now, with that quantum strangeness in mind, let's take a step back, all the way to the inception of our physical universe. The big bang theory is the prevailing cosmological model for how the universe began. It says that the universe started with a single mass in a very high-density state that then expanded to create our entire universe and everything in it (including the physical form of you and me). After the big bang, it took nine billion years for our planet Earth to coalesce, and then more than four billion more years for life-forms to inhabit it. And here we are.

The quantum and big bang theories viewed together pose an in-triguing question: Which existed first? Life or the universe that contains it? For any single particle to exist—including those that make up that original mass, the expanding gases, and the original Earth, every particle of oxygen in its atmosphere, and every drop of water in its rivers—some kind of life was needed to observe it into existence. Unless the laws of physics as we know them did not apply from the point of the Big Bang until life appeared in its physical form, then life existed before the physical world did.[3]

Since the Big Bang, time has been one of the most persistent properties—though illusive—of this physical world. And this is where our third theory kicks in. Einstein's theory of relativity delivers another mind-bending scientific conclusion: that all of time already exists in a four-dimensional structure called space-time. As discussed previously, the relativity of time means that you and I can have a very different concept of time depending on our speed, location, vantage point, and various other parameters. Accordingly, the absence of absolute time makes each of our perceptions of the beginning and end of any specific event different.

When you put together those three monumental theories—quantum, Big Bang, and relativity—you find that life, which encompasses the continuum of all possible observers, came first. That means it doesn't abide by the rules and principles of the physical world it observed into existence. And that puts us face to face with some very difficult questions: How, then, does a life end? An end is a point in time. When does it begin? According to whose time? Yours? Mine? When all of time always exists, then which life came first? Mine or Ali's? Who died first? Ali or me? What is "first," "last," "before" or "after" if all of time equally exists always? There's only one answer:

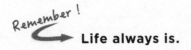

Remember! Life always is.

Our physical form is subject to the limitations of the physical universe, but in Einstein's conception of that universe, a slice of its space-time can include Ali dying along with my own birth. The real observer of that slice has to exist outside the limits of space-time, as part of the life that precedes the universe itself. The real you and the real me, outside our physical forms, living our long life continuum, transcend the arrow of time.

Now that's a tough one. Perhaps I need to leave you to think about these concepts for a while. Take as much time as you need, but please remember: the physical self is an illusion; life is not the body that is subject to the limitations of space-time. When you think of yourself as the observer, think of the real you, not the physical form that represents you. That's what life really is.

I believe Ali's physical form was a descendant of mine, but Ali's life isn't. His life always existed, and so has mine, outside the boundaries of space-time. In my definition, death is the end of our physical form, but it is not the opposite of life. Death is the opposite of birth. Birth and death are the portals through which we come in and go out of this physical form, but life is independent of all that's physical. Life *observes* the physical. It resides outside of it, where there is no before or after. In this lies the source of the peace I've experienced since Ali's departure. I know we will meet again.

Remember! ➤ **While our physical form decays, we never really die.**

What's after Death

Death scares us because we're comfortable with the familiarity of this life. We feel safe here, a bit like when we were in our mother's womb. Back then it was warm, there was free food, no time pressure, no taxes, and everything was chill. Imagine if someone had shown up then and told you that you were going to suffer the pain of a process called labor, which would squeeze you out of your familiar home, and once you were out, the food and oxygen supply would be disconnected and the peaceful darkness would be replaced by glaring lights. You would have said, "Hey, I am not signing up for this. I like it here. This is as good as it'll ever get."

But was it really? Would you want to go back now? Don't you think it is a bit better out here? So apply that thought to the next transition. We go through life with all of its ups and downs until someone tells us that at a point in time we will have to go through the pain of a process called dying and get kicked out of this home. Not surprisingly, we would react exactly the same way: "I will not sign up for this. I like it here. This is as good as it'll ever get."

If we could know in advance that it'll be fine after we die, it wouldn't matter so much that we do, would it?

Millions of near-death experiences have been documented in the United States alone. Simply put, these are cases of people who experienced death and came back. Most tell a very positive story. One of the most fascinating is the experience shared by Anita Moorjani, author of *Dying to Be Me*. In her TED Talk she said:

I shouldn't be alive today. I should have died on February second, 2006. I was dying from a late-stage lymphoma, which I had struggled with for four years. That morning I went into a coma, and the doctors said these were my final hours because my organs had shut down. Even though my eyes were closed I was aware of everything. I was aware of my husband who was distressed by my side, holding my hand. I was aware of everything the doctors were doing. It felt as though I had a 360-degree peripheral vision. I could see everything, but not just in the room. I was aware of my physical body. I could see it lying there on that hospital bed, but I was no longer attached to it. It felt as though I could be everywhere at the same time. Wherever I put my awareness, there I was. I was aware of my brother in India rushing to get on a plane to come and see me.

I also became aware of my father and my best friend, both of whom had died. I became aware of their presence with me, as though they were guiding me. One thing that I felt in this amazing expansive state was that I understood everything. I understood that I was much greater; in fact, all of us are much greater and more powerful than we realize when we're in our physical bodies. I also felt as if I was connected to everybody. I felt I could feel what they were feeling, but at the same time I didn't get emotionally sucked into the drama. At first I didn't want to go back into my sick and dying body. I was a burden on my family and I was suffering, but in the next instant it felt as if I completely understood that, now that I knew what I knew, that if I chose to go back, my body will heal very quickly.

To the surprise of her doctors, Anita woke up from her coma. The hospital records show that within five days the tumors in her body had shrunk by 70 percent, and after five weeks she was released from the hospital to go home—completely cancer-free.[4]

Most records of near-death experiences are similar: people go through a tunnel, see light, and meet their loved ones. They see gardens and rivers where everything is peaceful, full of love, and drama-free.

Many say that near-death experiences are just a biological brain response related to the process of dying. To that I say, *So what?* What difference does it make if that's really what life after death looks like or if it's our brains still refusing to shut up even as they're dying? So what if this is the last chatter they produce, just like when Windows crashes and displays that beautiful blue screen before it finally shuts down. Either way, it's not the least bit scary (not to mention super cool).

I personally experienced the wonderful ride of a near-death experience when a minor operation went wrong for me at a similar age to Ali's when he died. I saw that light, rushed through the tunnel, and found the calmness and peace so common in near-death experiences. To tell you the truth, it was fabulous. It was so much fun, that I wouldn't mind going for another ride.

But until then, let me focus on the other inevitable state: life.

Our mortality, ironically, is a life coach. Before you die, you might as well live a happy life. So let's learn how to find happiness despite death—or even *because* of death.

Death and Happiness

The path to joy is to see life for what it's worth. Ali's death put my life in perspective. While he was leaving, I imagine that he turned back and gave me one last gift—by taking something with him. He stripped away life's deception. He left life bare. Things that seemed to be important were exposed as worthless, and the true essence of what life is all about was left to shine through.

Death anchors us in the truth. It's the signpost that takes away all illusions. Believe that you have control, and death will shatter your

Life's Rough Patches
as Seen in Comparison to the Truth
of Our Mortality

The insignificance The diminishment
of the events of expectations

illusions. Associate too much with the physical world, and death will remind you that everything physical falls apart. Be proud of your knowledge, and the mystery of death will keep you puzzled. Try to slow down the decay of life, and death will crush your sense of time. When you accept the reality of death, there will be nothing left to fear. Only then will you finally live illusion-free. Without illusions, you can rise above thought to the highest level of joy.

Like every other truth:

Remember!

➥ Accepting death will set you free, but first it will really piss you off.

The World's Best Life Coach

Islamic culture counsels, "When you seek a teacher, look no further than death." If we really attend to death, listen to it, and talk about it, instead of trying as hard as possible to pretend it doesn't exist, it can teach us three lessons—not how to die but how to live a worthy and fulfilling life.

Lesson One: Death Is Inevitable

However unwelcome it may be, death will win eventually, so what's the point in spending your life fighting it? The best generals never fight a war they will lose—they focus their thoughts and energy on what they *can* affect. The first lesson death teaches is to accept it.

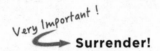

Very Important !

↳ **Surrender!**

Lesson Two: Life Is Now

The start and end of your life are like the covers of a book. Significant as they might feel, neither event really matters as much as the story that fills the pages in between.

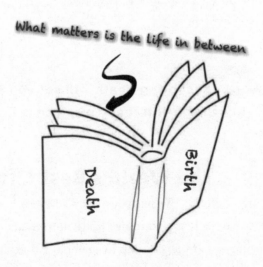

How would you live if you knew that today was your last day? More important, why are you not living that way today when you know that it may be your last?

If you knew for certain this was going to be your very last meal, would you be upset that the waiter was not friendly? Or would you slowly savor and enjoy every bite? If this were your last traffic jam, would you spend the time cursing? Or would you wish it took longer? Would you honk the horn in anger, or would you switch on the radio and listen to your favorite song one last time? Why does it have to be the last time for you to choose to relish the moment?

Remember!

→ **Live this moment as if it were your last.**

Once, after Ali left, when we were flipping through his beautiful pictures, Nibal showed me some baby pictures of him and said, "He was such a peaceful baby. Never cried and never complained. That baby visited for a while, and then he left. That form was gone forever, and then came the toddler. He was so curious and happy, and then he left; the toddler too never returned. The child who was so decent and pleasant visited, and then he left for the loving, giving boy to come, followed by the calm, knowledgeable teenager, and finally the handsome, wise man. Now that person too has left. I enjoyed knowing them all and miss every one of them, but each one sooner or later had to leave."

Every day a version of you and everyone you love dies. They leave and never return. Please don't let *any* of them pass unappreciated. We rush through life and delay living it. We keep adding to our bucket list, forgetting that the time to live that list may never come. Life is one long bucket list. Live it while you still can.

Very Important !

→ **Live before you die.**

Lesson Three: Life Is a Rental

Understand that when the time finally comes, you will leave everything behind: material wealth, the people you love, and everything you hold so dear.

This raises the most important question of all: Why do we hold on so tight when sooner or later it will all be gone? If you know for certain that you will leave all your money behind, why are you so absorbed in collecting more than you need? If someone else will sooner or later take your job, why are you so afraid to lose it? Why do we accumulate material possessions that we don't need today when tomorrow may never come? Once again, it takes simple math to understand what's wrong with what we do and what we should do differently.

Life is a zero-sum game: we come into it with nothing and leave it with nothing. For this to be mathematically correct, everything we are ever given has to someday be taken away.

Remember !

→ **Nothing can be gained that will not eventually be lost.**

You can read this with sadness, or you can let the truth set you free. My whole life and all that I ever called mine is, essentially, a **rental**. I enjoy it fully while I'm the tenant, but sooner or later, I'll happily hand it over to another. I find freedom in that. **If nothing is mine, then nothing can be lost.** So I let things come and go, and I experience them while they last. I love them wholeheartedly, enjoy them, and make them feel how much I appreciate them until it's time to move on and let them get on with their own life.

When I finally learned to let go and let everything flow, I seemed to end up with more—a counterintuitive fact that has an elegant geometry to it. Whenever something moves out of my life, space is created for new experiences to move in. Letting go makes my life richer. It's like the sharing economy: you can get driven around in the nicest cars—without ever owning one. So:

Remember! ⟶ **Rent a full and happy life.**

A rental life keeps me hopeful because I realize that bad times end too. Times of sadness, sickness, loss, or deprivation will pass. The scars we sustain, the weaknesses we attain, are temporary. Nothing's here to stay.

To die is to leave everything behind. Grammatically speaking, the verb *to die* never takes an object—dying can't be done *to* me—only a subject: *I die.* I'm not afraid to die because **I choose to die on my own terms.** I choose to let go of my attachment to all physical possessions before they are forced away from me. I choose to rent every experience that comes my way, enjoy it fully, but let nothing own me. When we learn to let go, we learn to die before we die. All of life becomes ours to enjoy, but not to keep. We find a life that's rich with variety and free of fear. Life becomes our focus. We stop thinking of the time we will rest in peace, and only then do we learn to *live in peace.*

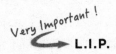

Very Important! ⟶ **L.I.P.**

The Game

When Ali died, I really wrestled to understand what life was all about. Writing helped me navigate the maze inside my head. As the puzzle pieces—around the Illusion of Self, Knowledge, Time, Thought, and Control, as well as the truth about death—started to settle in place, the picture became clearer. Finally, it all came together into what I now consider the core of my life's philosophy.

If the real you is neither your body nor your thoughts, then it's hard to resist the temptation to consider how the real you connects to the physical copy of you and commands it to roam this world we live in. The easiest way for me to imagine this relationship is to visualize how a player controls an avatar in a first-person action video game. In video games, *first-person* refers to a graphical perspective that is rendered from the viewpoint of the player's character, as though looking out at the game world through the avatar's eyes. In such a game, the player uses a game controller to command the character's every move.

For years, Ali and I played video games together. Our favorite was Halo, a game where we would play as "Master Chief." Throughout the game the characters we played would be swarmed by thousands of aliens, crawlers, and monsters. We would be attacked, shot at, thrown from high places, blown up, stormed by war vehicles, stabbed, and left to die. The terrain around us was harsh volcanic lava or slippery slopes. Danger threatened us from every angle, and everything we saw in the rugged landscape was out to hurt us. Master Chief, however, was a seasoned veteran. At our command he would rush into the game wherever the action was hottest, shoot his enemies, and move ahead. He would get bashed and battered. He would get wounded and fall in the dirt, only to get up again and get on with the game.

Ali and I would spend the time communicating strategy, complimenting each other on every good move, and occasionally teasing each other over the poor ones. We would pay full attention to every move and engage as if the attacks were real. The big-screen TV, the exquisite graphics, the dramatic music, and the realistic surround-sound effects of bullets zooming by and loud explosions shaking the room made it all feel very, very real. Fully absorbed in the game, we would play for hours and lose track of the "real" world until it was time to stop, and then, regardless of how harsh the game may have been, we would put the controller down and say, **"Wow, that was fun!"**

Fun?! This was *brutal*, you might say, if all you looked at was the screen. You'd seen a man being beaten, blown up, shot, attacked, and hurt from every conceivable angle. The whole world was against you. It was a massacre, you'd say. How could you think of this as fun?

The answer is simple: it wasn't *us* suffering. None of the beating, bashing, or shooting affected either of us. Winning was irrelevant. It was all about the gameplay. I was sitting on the sofa with my wonderful son, and it really was fun.

Now please consider the following: How different is your life on this earth from a video game? If your physical form—the avatar you use to navigate the physical world—is not the real you, then what difference does it make if you face a few challenges on the way? If the world deprives you every now and then, what impact does that have on the real you, the you on the sofa holding the controller? Regardless of how immersed we get in the game of life, we get through it. We live through ups and downs, some gifts and some losses, but none of it matters because when we focus on the gameplay every experience feels new, and it's all fun. Now that is a true gamer's point of view.

Serious gamers, you should note, always set the difficulty level of their games to high. When Ali played alone, he used to set Halo to "legendary," the most difficult setting possible. He would turn it down to "hard" only when he played with me.

When games are too easy, there's no challenge. It's slow and boring, and there's no fun in that. Only when the game gets harder do we engage, learn, and develop new skills. The best players out there get bashed, and then they learn, adjust, and jump right back in. Strange as it sounds, the harder the game gets, the more fun it becomes.

Take the hard parts of life with a smile. It's just the way the game is designed. Don't be fooled and buy into the sound effects. Don't let the fake explosions scare you off. During the game, Ali would always run his avatar to wherever the loud noises and smoke were coming from. I would ask him where he was going, and he'd reply that that's where the action must be. **That's where the best parts of the game are.**

Let's talk about levels. A level in a video game is the total space available to the player while trying to complete a discrete objective. When you reach the end of a level, you go through a portal of some sort; the game console briefly goes dark as it loads all the details of whatever comes next, and when the screen lights up again you're in a totally new environment, a new level. You might leave the urban battlefield and find yourself in a jungle. The new level changes the feel of the gameplay. The jungle may slow down your movements or obscure your vision, adding a challenge and more fun to the game.

Through each level you acquire new skills and develop your knowledge of the gameplay as you strive to achieve targets. When you achieve the "level purpose," there's no point in staying at that level anymore. You take very little, if anything, of what you've collected and proceed to face the challenges of the next level.

Intriguingly similar to life, no?

While the purpose of life may be a bit trickier to figure out compared to the target of a certain level in a video game, the process looks really similar. We come to this level of life from a previous level we know nothing about, through a portal called birth, and we leave it for a level we don't yet know, through a portal we call death. Could it be that this life is just one level in our larger game?

Most religious and spiritual teachings seem to believe this to be true. They tell us that death is just a portal to another life and that we never really die—it's just our physical form that dies. When you leave this level, you take nothing with you, although your good deeds in this level may position you better for the next. Some religions believe that if you fail to acquire the skills you needed to as you rushed through the game, you go back and replay the level through reincarnation.

Let me push the analogy even further and tell you about cheats and shortcuts. I've mentioned that Ali was a serious gamer. While I struggled with the controls and with translating the landscape into images my middle-aged brain could understand, he would run through the game as if he were using his real eyes and legs. When we played together, he was always a few steps ahead of me as I rushed to catch up. He would race through the uninteresting parts of the level and linger in the fun parts, enjoying everything the game had to offer.

Occasionally Ali would take a turn and stop in front of a tree or a brick wall. After standing there for a moment and looking back to see where I was, he would run right through it, revealing a shortcut, a cheat that took him directly to the next level. He would then put his controller down and lovingly say, "Don't worry, Papa, I'll wait for you here." Sometimes I would have to plow through the whole level to get to the end and go through the portal to catch up to him, and sometimes I would figure out how to find the same shortcut. He would always be there when I got to him. He would smile, give me a high five, and say, "I'm proud of you, Papa," and then we'd go on to explore the next level of the game together.

Ali had a full life. He enjoyed the best parts of this level—this life—with friends, music, and a lot of love. He was always happy. While I have no scientific evidence to prove it, I believe that on July 2, 2014, he found a shortcut. At 4:11 a.m., while Nibal and I were sitting worried outside the ICU, we felt an overwhelming surge of positive energy that gave us a feeling of relief. His uncle texted us from thousands of miles away to say he felt the same.

Seconds later a doctor rushed out in a panic. He called in other doctors, who rushed around frantically for a while. We, on the other hand, sat there calmly. We knew it was all okay. Though they later came out and claimed Ali was stable again, it was clear in my heart that he had found his shortcut. He had taken one loving look at both of us and walked right through, saying, "Don't worry, Papa, I'll wait for you here."

One day, when my work here is done, I too will get to the end of this level. We all will. Do your thing, *ya habibi*. I will catch up to you soon.

Don't you get it? It is all just a game. So play, live, learn, and:

Very Important!

Have fun!

Ali's Last Wish

As if he knew he was leaving, for his last couple of months Ali asked almost everyone he met, "What happens to us when we die?" Typical of Ali, he would ask the question and listen attentively to the answer. Then he would ask clarifying questions, listen some more, nod his head, and say, "That's so interesting!" He received an extremely diverse set of answers. In one of his very last conversations, just a few days before he left, he finally shared his own views with a friend. He said, "I guess we will only know when we get there, but I am optimistic. All I want when I get to the other side is to go to the highest place and see the face of the one who made this awesome universe."

Even as he was leaving, he took the time to leave a message. He told us that he found his peace days before he left. Live on in peace, my wonderful friend, but please answer me one last question: Did your wish come true? Is there a Game Designer? Did someone really make all of this, or did we make that someone up?

This makes for one more truth to discuss. Don't stop now. Please read on.

Chapter Fourteen

Who Made Who?

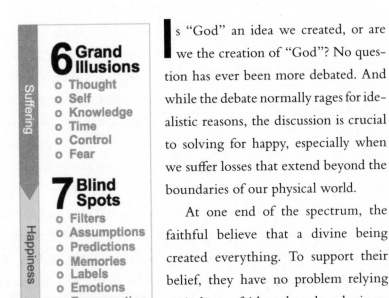

6 Grand Illusions
- o Thought
- o Self
- o Knowledge
- o Time
- o Control
- o Fear

7 Blind Spots
- o Filters
- o Assumptions
- o Predictions
- o Memories
- o Labels
- o Emotions
- o Exaggeration

5 Ultimate Truths
- Now
- Change
- o Love
- o Death
- Design

Suffering

Happiness

You Are Here

Is "God" an idea we created, or are we the creation of "God"? No question has ever been more debated. And while the debate normally rages for idealistic reasons, the discussion is crucial to solving for happy, especially when we suffer losses that extend beyond the boundaries of our physical world.

At one end of the spectrum, the faithful believe that a divine being created everything. To support their belief, they have no problem relying entirely on faith rather than logic or science. At the opposite extreme, materialists posit that no such entity exists and that a repetition of random events over an unfathomable amount of time is the creator. The Big Bang started things off, and then evolution and natural selection carried us to the current day. There appears to be very little basis for common ground. We all believe in something, but no one seems to agree what the core of the debate is about.

What we mean when we say "God" varies dramatically across cultural, spiritual, and religious backgrounds. So a lot of the arguments end up being misunderstandings rather than fundamental disagreements. But our inquiry into happiness is a search for truth, so we *are* interested in the fundamentals—particularly in understanding why life seems to randomly surprise us, even mistreat us, and miss our expectations. The question with the most bearing on our search for happiness is this: *Is life and our universe the product of randomness or design?*

The question of design is strongly correlated to the question of a designer—a God. When I set out to write about the grand truths, I was explicitly advised to avoid this topic. Debating God is a sure path to polarize readers, and no good can come of that. But the topic kept coming up because the concept of the grand design played a pivotal role in enabling me to face the loss of Ali and sustain my state of joy. Indeed, my model for joy would've been missing a critical pillar if I believed that my loss was just a random roll of the dice.

The idea of the grand design suggests that every tiny movement in our universe follows a meticulously intricate pattern, that nothing is random. This sense of pattern helps me understand that the long lines in the supermarket are not because it's my "unlucky" day but because the laws of supply and demand drove higher traffic to that location at that time. Design means that the tsunami in Asia wasn't the result of an angry or absent God; it means that the movement of a tectonic plate led to an earthquake under the ocean. When we eliminate the self-referential stories our brains create about why events happen, we realize that everything happens as part of a highly synchronized universe where specific equations (though not always known to us) always apply. This simple adherence to the truth can be a life changer because it has the potential to solve your Happiness Equation once and for all.

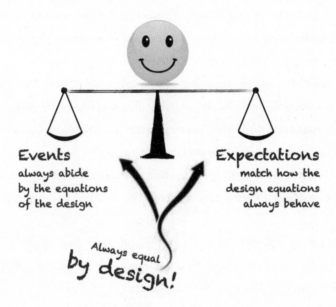

Take a simple example: we know that the opposite poles of a magnet attract, and because we know there's a physical law that accurately governs that behavior, it would be foolish to expect otherwise or to get upset when it happens. Similarly, I know that, at some level, events always match what I ought to reasonably expect, and while I don't always *like* the way events unfold, I would be naïve to expect that they would unfold in a different way. In that, I find peace. As I try to influence the events with my input into the equations, I understand that I'm but one in a million parameters that affect the path of life.

There's more to design than calibrating our expectations. Belief in design implies a belief in the existence of a designer. That has a significant impact on our state of joy. If you can empathize with my loss of Ali, you'll probably see that believing in the existence of a designer—

believing that we are part of something bigger than this physical world and that therefore Ali is okay—is for me a more comforting story, regardless of its scientific validity, than believing that he just vanished into nothing. Believing in such a "fairy tale" helps ease my pain a bit. But what helps me even more is my strong conviction that it's not just a fairy tale.

What if I could prove, using solid mathematics, how true the concept of a designer really is? It's possible. For me, the mathematical proof is what made the difference between simply finding *comfort* in the story and regaining my true *happiness* after Ali left.

So let me share with you the concept of a designer, not from the mind of a religious adherent but from the mind of an analytical engineer.

As always, remember, this is just *my* point of view. Keep what you like, ignore what you don't, but please don't rest until you find a path that you can call yours—your own truth.

We'll be attempting to answer a question that has puzzled humanity for ages. So give it a few pages before the various puzzle pieces of logic come together to form one coherent frame.

Ready?

The Problem Statement

It's as true in engineering as it is in business as it is in life: the most crucial step on the path to finding an answer depends on the question itself. **If we don't know what we're solving for, then whatever answer we may find will be irrelevant.** So let's accurately define the problem statement.

To avoid confusion, and to get beyond the centuries of heated debate associated with this subject, let's avoid common terms such as *God, Creator, Divine Spirit, Higher Power, Universal Consciousness*, or even *The Force*. Religious and spiritual institutions have blurred the true meaning behind those terms and often shaped them to fit an agenda. Instead, I'll use *the designer*, a term that strips our question down to its purest core.

Layering

Another important step is to find the simplest possible form of the question by removing the layers of ostensibly related questions. When we solve the core question first, it'll be easier to answer questions that flow from it.

In a popular stand-up routine, comedian George Carlin joked about many of the layered issues of religion and God:

Religion has actually convinced people that there is an *invisible man* living in the sky who watches everything you do every minute of every day. He has a *list of 10 things* he does not want you to do. And if you do any of them he has *a special place full of fire* and torture where he will send you to suffer forever and ever *till the end of time.* . . . But he loves you! . . . And *he needs money*. He's all-powerful, all-perfect, all-knowing and all-wise but somehow just can't handle money. I tried to believe in God but the older you get the more you realize, something is wrong here. *War, disease, death, destruction, hunger, filth, poverty, torture, crime, and corruption*. This is not good work. Stuff like this does not belong on the resume of a supreme being. This is the kind of stuff you'd expect from an office temp with a bad attitude. If there is a God then most people wouldn't disagree that he is at least incompetent or maybe, just maybe *doesn't give a [censored]*.[1]

The issues raised here are significant, valid, and worthy of discussion. I'm sure such issues have crossed your mind too. They all, however, are good examples of **layering**.

It's easier to untangle this .. than that!

When additional layers get entangled with the core problem, our thought process becomes too multi-threaded, the conversation strays from the shortest path, we get frustrated, and the problem becomes much harder to solve. A more efficient approach is to strip the design/designer question of all its distracting layers.

Carlin addresses specific claims, interpretations, and fables of the religious institutions, which claim ownership of the "God" brand. I agree that many such fables are ridiculous and provoking, but they're irrelevant to our conversation here. Think of it this way: if someone makes up a ridiculous story that Facebook came into existence because a lightning bolt struck Mark Zuckerberg's computer, that wouldn't make you assume that Zuckerberg doesn't exist, right? Disagreements with the story of judgment, an eternity in hell, evil acts of mankind, ruthless natural disasters, or other actions attributed to that God are—once again—irrelevant. They are similar to your disagreement with a specific political party: you know that your disagreement doesn't serve as a proof that the party does not exist.

I'll try to sidestep the countless layers of concerns and focus on our

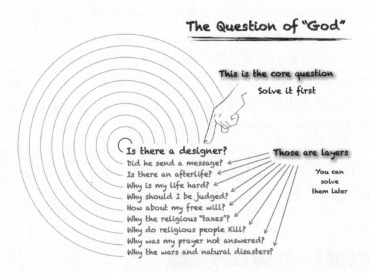

The Question of "God"

This is the core question
Solve it first

Is there a designer?
Did he send a message?
Is there an afterlife?
Why is my life hard?
Why should I be judged?
How about my free will?
Why the religious "taxes"?
Why do religious people kill?
Why was my prayer not answered?
Why the wars and natural disasters?

Those are layers
You can solve them later

logic stream until we arrive at an answer to the core question. **Those layers will not be ignored**, just kept aside in order to address the problem one layer at a time. For now, please assume no association of the "prospective" designer with any religion, fairy tale, attributed action, or presumed instructions. Let's agree to solve one problem first: *Is our universe the result of randomness or intelligent design?*

If we end up reaching the conclusion that there is no designer, then all the layered questions are pointless anyway. If, on the other hand, we reach the conclusion that there likely is one, then you can start to plug the other questions on top of that foundation one by one. Did the designer create? Was there a message? And so on. So let's begin. As you might have guessed, my background leads me to start from a certain point of view.

It's a Math Problem

I was born a Muslim. As in most religions, Muslim scholars have focused for centuries on mechanical practices: do *this* and don't do *that.* They've ignored the core of Islam's spirituality, and they've even openly discouraged people from seeking their own answers. At sixteen, I rebelled and decided to revisit the hypothesis. I declared (if only to myself) that I was agnostic, and I went on a quest to seek my answer. I stripped away all the layers of urban legends, fables, gospels, and emotions. What was I left with? Mathematics! So I started to crunch the numbers and decipher the facts surrounding intelligent design. In place of all the murky old confusions, I found two terms in fundamental— and solvable—opposition: *absence* and *presence.*

Absence versus Presence

As an agnostic, I found it easier to take the atheist side of the debate and pose the question this way: How can you *prove* that there is a designer? The creationist background in me would jump in and blabber the answers I was taught—spiritual stories and old scriptures that don't prove a thing. My internal debate was not leading anywhere until I realized that **the question was incomplete**, and hence the answers were always insufficient. A balanced, agnostic question should ask for proof that there *is* a designer **as well a proof that there isn't**. This framing would put the burden of arriving at a proof equally on both sides of the discussion. I was surprised that the atheist side of the debate avoided framing the question in such a way, but once I asked it, the reason became clear:

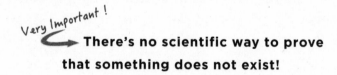

Very Important !

There's no scientific way to prove that something does not exist!

From a scientific method viewpoint, it's not possible to prove a negative. This might sound simpler than it actually is. But consider this: Obviously it's possible to prove that something—say, monkeys—exists. All you need for the proof is to find one. Locate a monkey—presto, monkeys exist, and you have proof. But it's impossible to prove that an imaginary creature—say, plunkeys—*don't*. You'd have to examine every possible scenario in every square millimeter of the universe to ensure it's free of plunkeys. Because our universe is so vast and complex, this task is impossible. Furthermore, because of the limitations of our senses, proving a negative will always be inconclusive. If plunkeys were infinitesimally tiny, we wouldn't find them until our instruments were sophisticated enough, and if they were larger than our entire known universe, we wouldn't be able to observe one, perhaps ever.

But here's the catch:

Remember!

→ Absence of a proof that something exists does not prove that it doesn't.

Various historic examples demonstrate that we've repeatedly missed fundamental components of our universe. We've missed almost all of it, as a matter of fact, for ages. My favorite example is how we looked up into space for millennia and assumed that stars and planets floated in nothing, in a vacuum, "empty space," while in fact everything is entirely submerged in dark matter. Was there ever a way to prove that this fundamental universal component didn't exist? No, never! There was only a way, back in the 1960s, to finally prove that it does, and even now we still don't *see* dark matter. We can only prove its existence by observing cosmological behaviors that refer to its presence.

Discovering the majority of the matter that makes up our universe after decades of confusing it for a vacuum should force us to ponder what else we can't observe even though it exists. Which suggests that we have to look at existence itself in a different way:

Very Important !

→ Absence of a proof that something does not exist should be seen as a *probability* that it does.

And that's where we need to start using numbers. (As I said, it's all math!)

A *probability*, in this sense, measures the likelihood of something's existence, regardless of how small. Even though you and I can be quite sure that plunkeys don't exist—I made them up—there's still a tiny probability that they do. Since I started thinking this way, the word *probability* started to pop up in every part of my investigation. And that led me to an important distinction that made the problem statement clear.

Has versus Could

The debate around the question of grand design lies firmly within the Illusion of Knowledge. One side of the debate strongly believes in a divine entity capable of intelligent design, and the other equally strongly believes in randomness. They both "know" they're right.

Unfortunately, **both are wrong!** No one can prove conclusively for or against either view. And in math, the absence of conclusive answers, the main premise of our conversation, should turn into a simple matter of probability, a question of which side is *more likely* to be true.

This small shift in perspective frames the problem statement appropriately. It becomes:

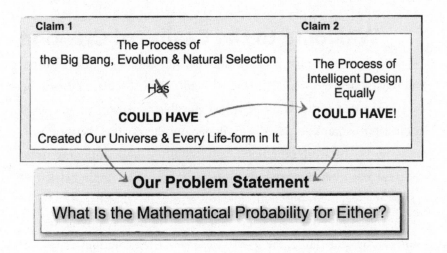

Claim 1	Claim 2
The Process of the Big Bang, Evolution & Natural Selection ~~Has~~ **COULD HAVE** Created Our Universe & Every Life-form in It	The Process of Intelligent Design Equally **COULD HAVE!**

➤ Our Problem Statement ↙

What Is the Mathematical Probability for Either?

Now that the problem statement is defined, we can start searching for the answer. All we need to do is calculate the probability of each side of the debate. No more intractable arguments. In fact, we can focus on just half of the question; the nature of probability theory allows us to solve for one side of the problem because once we know the numbers supporting that side, we can find the probability of the other side by subtracting from 100 percent. So let's crunch the numbers for the scientifically supported, randomness side.

Would you like to take a short break while I do the work? That might not be a bad idea. Come back fresh. We have some very large numbers ahead.

Welcome to the Casino of Creation

The materialist side has it that a sequence of random events, regulated and channeled by natural selection, is sufficient to create everything we know. Randomness creates every possible scenario (in the case of a dice, for example, 1, 2, 3, 4, 5, or 6), and then natural selection intervenes by discarding the first five possibilities and keeping the sixth. Natural selection doesn't reduce the number of attempts needed to arrive at a specific result; it just discards errors after they happen. Those errors in the case of creating a complex system, such as a living organism, can be quite large, but *given enough time*, the evolutionists believe, random trials can produce a result that matches our universe and every life-form in it. My mathematical brain totally agrees with this view. The equations are correct. Given *enough* attempts, any configuration, without exception, is possible.

But the fact that it's possible is not a proof that it's true. Imagining the sheer scale of creation regulated by natural selection in the real world is totally different from plugging numbers into an equation. Evolution *could* very well have created everything. The question is *What are the odds that it actually did?* How large are the numbers of random attempts that we're dealing with here? Let's start with a simple example of randomness to ease us into the math.

Imagine being told that you could earn $1 billion in a casino where you'll be given several boxes with a number of dice in each. Your task is simple: just roll the dice. If every single roll lands all dice on the 6, you join the billionaire club. Ready to play?

A Question of Luck

A roll of the dice best represents randomness. If you roll enough times, you will eventually yield every possible result (1, 2, 3, 4, 5, and 6). Sooner or later you're bound to succeed—but exactly how much *later* is later? That depends entirely on how complex is the result you're attempting to achieve.

First, roll just one die and try to get a 6. No mystery here—you'll roll a 6 about once every six attempts, on average. If you're a very lucky person, it may happen earlier, and if you are not, it may take longer, but it is reasonable to expect such odds in general. Easy!

Now aim for a slightly more complex result. Roll two dice and aim for a double 6. Things start to get tricky but still are not too hard. You just need to get a bit luckier. The chances for each of the dice you roll is still 1 in 6, but your chances for both to do it at the same time do not double because you doubled the number of dice; it gets squared. It's not 1 in 12, but 1 in 36.

This trend continues, and it does not take long for your chances to fade away as the number of dice—the complexity of the system—increases. If you roll three dice, you will need, on average, 216 attempts to get a triple 6, and if you throw 10 dice, just 10, your chances become a dismal, 1 in 60 million.

Rolling ten dice seems like an easy task, but if you were betting your happiness on it, would you play the game with just ten dice? What would your chances be? Please think about that for a minute before you read on. Would you make that bet?

Now compare this process of rolling just ten dice, a complex system, to the complexity of creating our entire universe, or even one living creature. It's not hard to see that the odds of that are equal to the odds of rolling millions, no, trillions of trillions of dice. Would you bet on that?

There's Just Not Enough Luck

Our universe's complexity goes beyond human comprehension and certainly exceeds my math skills to simulate. It would be more manageable to assess the probability of just one small part of it, a single scene. Or even easier, a novel that describes that scene. What are the chances of that being written purely randomly?

Let's borrow Brian Greene's famous example from his book *The Fabric of the Cosmos* to demonstrate the barely comprehensible complexity of the universe as illustrated with a novel. *War and Peace* is the epic novel by Leo Tolstoy that describes the events surrounding the French invasion of Russia as seen through the eyes of five Russian families. It takes him more than 560,000 words to capture this tiny sliver of our complex universe. Tolstoy did not *create* the events of the era, nor did he *create* the five families, or France, or Russia, or Napoleon and his troops, or the snow with which they struggled. He just organized his words in an orderly manner to describe those events. Achieving that *highly simplified* version of our universe, however, is still very unlikely to happen randomly. You would have to put the letters in the correct order to make up hundreds of thousands of words. Those in turn need to be put in the correct order to make up tens of thousands of sentences, thousands of paragraphs, and eventually hundreds of pages. We can calculate all those probabilities starting with the simplest part of the task in an experiment you can do yourself, sorting the pages.

Simply buy a copy of *War and Peace*, take the pages apart (there are 693 double-sided pages in some editions), and throw them up in the air, allowing them to land randomly. Assume that a miracle of physics will get them to land in one stack (and not scatter all over the room), and ask yourself this: "What is the likelihood that they will land in order with page 1 on top, followed by 2, then 3, all the way to the end?"

There is only one possibility for the pages to land correctly and a very large number of possibilities for them to land out of order. To be specific, there is exactly $10^{1,878}$ (that is, 1 followed by 1,878 zeros) possibilities for the pages to land.[2] Only *one* of those would be correctly sorted.

Those kinds of shockingly large numbers are left out of discussions of evolution and intelligent design. But now that you see the numbers, your bets become more informed. If a slot machine needed to be fed so many trillions of coins for every jackpot (which in this case is one orderly copy of *War and Peace*, nothing more), how many players do you expect will be lining up to play? Would you?

What is the only way to get those pages back into an ordered pile, may I ask? **Intervention.** Someone needs to pick those pages up and do the *intelligent* work needed to create a readable novel.

Let's keep going. Take a closer look, from pages to sentences. Let's give a monkey (call him Randy) a typewriter and teach him to push the keys. Randy is *not* an author, so he'll just produce a stream of random letters. Give Randy an infinite supply of paper and an eternity of time. Writing a legendary novel is no easy task, so let's check the effectiveness of our random monkey with a simple sentence first.

This short sentence can be produced by random keystrokes

This sentence consists of 56 positions. Each can be filled by either a letter or a space to be chosen randomly from an alphabet of 26 letters + 1 space bar. Every time Randy types 56 keystrokes at random, we will check to see if he has written one sentence correctly. Easy, right? Not at all.

Assuming that Randy is the fastest typist on earth, typing 220 words per minute, we can check once every 2.5 seconds.[3] Tedious, I know, but let's finish this quickly so that we can move on to the main task. How long do you think it will be until randomness produces a satisfactory result? Well, it will take a while: 143 million-trillion-trillion-trillion years, to be exact, during which you will have had to check 11.4 trillion-trillion-trillion-trillion-trillion-trillion wrong spellings.[4] Oh, by the way, that's approximately 2.5 billion-trillion-trillion-trillion-trillion-trillion-trillion multiples of the age of our planet Earth![5] Shocking for such a simple task, isn't it? So what about the main task, writing *War and Peace*?

Don't hold your breath. If Tolstoy wrote *War and Peace* using random keystrokes, assuming 6 letters per word on average, it would have taken him $27^{3,480,000}$ attempts (that is, 27 multiplied by 27, 3,480,000 times) to finish.[6] Randy's version would be without punctuation, and the entire book would be on one long line, which would make it much harder to read but, hey, cut the monkey some slack. Type that number into any powerful calculator and the intelligent machine will replace the value with something more understandable: *infinity*.

Waiting for that one orderly copy would take a very long time, equivalent to a mathematical never. *War and Peace* simply could not be written by randomness even if our random monkey could live to be trillions of orders of magnitude older than our universe. Expecting that this could happen without an intelligent author would be a mega bet, one that a mathematician would not recommend you take. I think you would agree.

Randy's "simple" task was to write a sentence, and then a novel. We didn't ask it to create the humans who lived the story, the land that was invaded, or any of the billions of living things that surrounded them but were not captured in the plot. What if we did? What if randomness started from a blank sheet not of paper but of void? What if Randy had to create the stars, the planets, the favorable environment to sustain life on Earth, life itself, with his random keystrokes? How many attempts would that have taken? Do you really believe that a monkey can do all of that?

When we see the numbers, our minds refuse to accept that a novel can be written by randomness alone. Why is it, then, that some of us are willing to accept that the complex world about which the story was written could be created by randomness? I wonder.

The Game Is Rigged

When Brian Greene used the image of throwing the pages of *War and Peace* in the air, his purpose was to illustrate an intriguing property of our universe, entropy, a property that I did not factor into the original calculation because it would've made the random monkey's task a whole lot harder.

Imagine if the typewriter was rigged so that it frequently erased the line that had just been written. A little disaster of this sort would make Randy's unachievable task even more impossible—and that's exactly how our universe operates.

Our world has a tendency toward chaos. Everything tends to become more disorganized along the direction of the arrow of time. Entropy, a measure of disorder, never decreases. You don't need a PhD in physics to see that even on the rare occasion when something comes together perfectly, the tendency of our world is to break it down, not make more of it. Chaos clearly is the way of the world we live in.

We can see a glass break, but never unbreak. We see wild plants cover abandoned places, but they never grow to form neatly cut hedges. Thousands of gas bubbles randomly rise from a fizzy drink, but never collect back into the bottle. The equation of chaos theory works against such luck. It makes it even harder for consecutive events in a complex system to happen in the right order.

And order is important. I already pointed out that the probability of rolling ten 6s using 10 dice is 1 in 60 million. Imagine if I asked you to get this event to happen 3 times in a row. Roll ten 6s, pick another batch of dice, roll ten 6s, and then pick a third batch and get ten more 6s. Improbable, isn't it? Now the odds for the survival of a flower are much more dismal. Even if a beautiful flower comes into existence due to random chance, it would still need a favorable environment to sustain its life. That environment did not include bees (fossil records show that those appeared later), but evolutionists dismiss that as a *small* detail, claiming that flowers didn't need bees for pollination then. Yet a flower surely would have needed any number of other things: rain, soil nutrients, sufficient sunlight, and so on. For each of those needs to exist, a complex roll of dice would have been needed. And for that, evolutionists will assume that all of these individual events kept happening randomly until, on a lucky day, they all coincided. There's nothing wrong with that story; it's plausible. But just remember how many rolls of ten dice in a row this would be equivalent to. Moreover, remember how often entropy rigs the game so that every time a flower shows up, our universe's tendency is to blow it away so that things become more disorganized.

An Overlooked Detail

Please don't get me wrong: *evolution is a scientific fact*. There's no point arguing that. We see evidence of it all around us. But often when evolution is discussed, a small detail is overlooked: a distinction between micro- and macro-evolution.

The recent discovery of biogenic graphite in western Greenland has led scientists to conclude that life on our planet began 3.7 billion years ago. Since then, scientists say, evolution took over and led to the diversity of biological organization at every level. Countless examples are cited to prove the theory of evolution: moths changing color in response to pollution, Darwin's finches changing beak and body sizes to adapt to new food sources, Italian wall lizards changing their head size to become vegetarians, and more. Each example is extraordinary and convincing. But there's an important limitation apparent in all of them: the change happens within the same species. This is known as *micro-evolution*. Those moths remained moths; the fifteen types of Darwin's finches are still birds; and the vegetarian lizards have not turned into cows. There is no evidence of a change of kind, known as *macro-evolution*, found anywhere in evolutionary literature. There has been no *observable* evidence of a change from a fish to an amphibian or a bird to a dinosaur. Scientists depend on a series of *historical* evidence that appear to be of chronological events, such as fossil records, and then build theories to explain them. But no one has ever seen an observable proof of those theories actually happening. Evolutionists will explain that these events take thousands of years to happen and that evolutionary theory is less than two hundred years old—that if we wait long enough we're bound to see such evidence. While that's a mathematically valid reason, it doesn't change the fact that without observation, the theory of evolution is at best a story. And do you know what we call stories that are not supported by observable evidence?

We call them, well, faith! It seems, as David Foster Wallace wrote:

Remember !

"Everybody worships. The only choice we get is what to worship."

Evolutionary adaptation *is* indisputable, but it's insufficient to explain the abundance of creation surrounding us. Could there be a more plausible story? To find it, we'd need to bust another common myth.

Evolution versus Design

Design is not the opposite of evolution. Evolution can very well be a method of deliberate design. You can see this in action when you look at the technological products we create. A process of consecutive iterations, releases, and versions makes every new product better than its predecessor. Yet each of the products is designed and engineered. None results from random strokes of luck.

Imagine if in a few million years, an alien species excavated a line of cars that had been "fossilized" and preserved in good condition. In this "*car*chaeological" site, they manage to find several specimens of a species named "Audi." If the aliens use today's evolutionary mentality to explain what they've found, they will be compelled to believe that Audis have evolved as a species.

Audi Quattro – The Missing Link
4-wheel Drive Propelled the Species Forward

I mean, look at them! They were really bland until 1980, when a clear evolutionary surge in the fossil of the Audi Quattro, "the missing link," occurred. It was the first time four-wheel-drive systems appeared. This made the Quattro much better than its ancestors, enabled it to win races against competitive species, and quickly made competing models extinct. It survived because it was the fittest. From then on, the Audi species lived through years of rapid evolution and finally arrived at much more sophisticated "*car*eatures," able to achieve much higher speeds with the R8 and to efficiently conquer the off-road environment with the Q7.

The whole species became a lot more reliable and so extended its life expectancy to enable every Audi to go for hundreds of thousands of miles. They became more aware of their surroundings through GPS navigation systems, cameras, and distance sensors, and they even started to understand spoken language and speak back. Through natural selection the older models were extinct, and only the fittest of all models survived. *They evolved!*

The aliens are right. It would be irrational to assume that Audis did not evolve. I think you would also agree, however, that it would be foolish to assume that they just randomly came into existence with no intelligent design and engineering intervention.

Please understand that the theories of evolution advocate that making a car totally randomly is *theoretically* possible. If we give it infinite time and endless trials and errors, the metals in a specific spot on earth may land in the exact correct configuration to become Audi cars. It is mathematically possible, but is it probable? Also please note that cars are much simpler creations than self-replicating living organisms. To simulate this much higher level of complexity you would need to imagine that randomness, say, a hurricane passing through a junkyard, leaves behind a perfectly designed, fully autonomous factory that produces perfectly designed cars. Is that mathematically possible? Sure—but is it probable? I guess not.

If our minds won't accept that a car can be produced by evolution alone, how can anyone bet on evolution as the creator of our much more complex universe?

When it comes to understanding evolution, it is mathematically correct to assume that a stroke of luck *could* have created everything, but to say that it *did* is not a wholly accurate statement—unless, as science would claim, the universe had **infinite time** to try. And that, unfortunately, is another *mega* myth.

There's Simply Not Enough Time

We did not have infinite time! The age of our universe since the Big Bang is approximately 13.7 billion years, and the age of Earth approaches 4.5 billion. We know that life—in its most primitive form—started 3.7 billion years ago. While this does sound like a long time, it's dismally short when compared to the task of creation. Our universe isn't even old enough for Randy to type a simple nine-word sentence, let alone create a house fly. But that shouldn't stop us from trying one more time, should it?

A faster monkey is all we need to finish the task *in time*. I mean, seriously, nine words in 4.5 billion years! There must be a monkey for the job. Randy is fired and replaced with a much faster monkey—call him Flash. Fast as he is, Flash does not come cheap. He asks for a very large banana as a reward and gets to work.

The task remains the same: 11.4 trillion-trillion-trillion-trillion-trillion-trillion sentences to be written for a fair probability of getting one right. To get that done, Flash needs to type at the blazing speed of 550 thousand-trillion-trillion-trillion-trillion-trillion words per minute.[7] Is that the rate of creation we observe on earth today? Or is that just another overlooked detail?

What if we multiply that need for fast creation by the number of living things roaming our planet. About 226,000 species are listed in the oceans alone,[8] and it's estimated that a further 2 million remain undiscovered. When we include all of life, approximately 8.74 million species populate the Earth.[9] Please don't be fooled by the familiar story of evolution and natural selection; the devil is in the mathematical details. If creating each of those variants is as simple as typing, and it isn't, then how many attempts do we need? How much faster should we witness those attempts to create in the world around us? I'm not even going to waste time doing the math. The point is clear.

Oh, and there's one more question.

Where Have All the Fossils Gone?

One of the notable differences between one of our nearest claimed ancestors, *Homo habilis*, and our species, *Homo sapiens*, is that our brains are three times larger. These two species appeared on the scene less than two million years apart. If we applied the same odds of our monkey typing his nine-word sentence to the evolutionary leap from *habilis* to *sapiens*, evolution would have needed to produce a number of *Homo habilis* that exceeds 20 billion-trillion-trillion-trillion-trillion-trillion multiples of our entire human population in order to create the first one of the *Homo sapiens* by chance.[10] The pivotal question then becomes: Where have those nearly infinite number of skeletons gone? They really shouldn't be too hard to find because on average there should be 137 million-trillion-trillion-trillion-trillion of them in every square meter of our planet's land mass.[11]

Have they all, without exception, decayed? Even if that somehow happened, why don't we see observable evidence of this pattern—the creation of 5.7 trillion-trillion-trillion-trillion humans per month—continuing today?[12] Are we so perfect that nature decided to stop us from evolving? What happened to everyone's math skills? Why are those simple calculations ignored?

Saying that randomness went through every possible configuration until it eventually found a "working" model to keep, then build upon, is not a plausible story. We don't need a fast monkey to go through all configurations; we just need a very, very lucky one. So lucky, in fact, that it will get things right on the first attempt every single time.

We Found Our Monkey

Imagine if a monkey existed that could roll ten 6s several times in a row, consistently beating the odds of 60 million to 1. If speed won't get the task done, then this monkey, Lucky Bastard, surely is our best bet. As we explain the task at hand, the monkey seems confused. We're asking it to roll 55 dice each with 27 faces and land them all on the correct face just to create a nine-word sentence. And if we generously assume that to be equivalent in complexity to, say, creating a fly, then it will have to repeat that stroke of luck 8.74 million times in a row to create every other living species in a highly choreographed sequence. Being a monkey, though, it agrees to give it a try, but instead of asking for a banana as a reward, it asks you to bet your sanity on its success. The monkey is willing to take that bet, but are you?

While I fully subscribe to the constant evolution of life as a law of nature, in my quest for an answer I had to stop and revisit the underlying premise. Evolution is a fact—but just as with Audi cars, consecutive releases of improved models, or perhaps just sheer variety, are just part of the design. The math is clear. Nothing is random.

Remember! **We are all part of a grand design.**

The Stuff We're Made Of

Enough about dice, Russian novels, and cars. Let's get to the bottom line and talk about the reality of what you and I are made of—proteins—which provides the very best example of just how impossibly complex everything we take for granted really is.

Proteins are generally made up of a bunch of amino acids strung together in a specific sequence. There are twenty different amino acids known to science today. The sequence in which these are strung determines how a protein molecule behaves.

Imagine a string of beads that come in twenty different colors. String together 30 green beads, followed by 1 white one, then another green and 12 blue beads, and you get a specific protein molecule that fits, say, in your muscles. String together 13 yellows, 22 reds, and 2 blacks, and you get another protein that works as an antibody. Each protein is a highly sophisticated machine that performs a specific function. Some will act as pumps, others will shape-shift to fight germs, and some will move like motors. Your body contains more than twenty thousand such machines. Many other living beings contain protein-made machines that exceed a hundred thousand types.

The most amazing thing about those strings is that they don't just dangle in a straight line; they fold themselves based on the sequence in which they are strung. And they keep folding until they find a stable "minimum energy configuration" to maintain the integrity of their structure. Like an origami creation, each fold needs to be done accurately in the correct sequence for the final work of art to come into shape. This perfection is very difficult for a protein to achieve without errors (known as "misfolds") because water molecules bang those tiny strings around as they fold, forcing them to move and jiggle.

Yep, it's that complex!!

Now forget about all of the evolutionary cycles needed to create your ancestors, and let's focus on just one of the twenty thousand proteins that make up our body. For that single building block to have existed it would have needed to be accurately strung in the exact sequence of amino acids that form its composition, and then it would have needed to fold correctly to find its functioning stable structure. How likely is that to happen due to randomness?

In 1969, Cyrus Levinthal noted that the protein molecule has an astronomical number of possible folds leading to its final structure. The Nobel Laureate Christian B. Anfinsen calculated that it would take 1,026 years for a single simple protein to form by randomly sampling all possible folds until it arrives at a stable structure. **This is more than a trillion times longer than the age of the universe!**

What would be the only way for that protein to fold correctly within the time it had for the task? You know the answer by now: intervention! The protein would have needed to know the primary sequence before it started to fold. Knowing how something is bound to be before it starts to be is what we call *design*! Like a documented origami work, the protein strand needed to be preprogrammed with the exact steps to follow to get the job done in time.

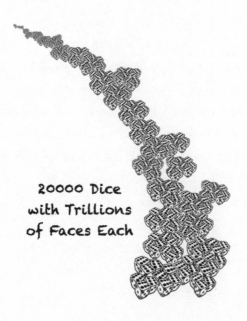

20000 Dice with Trillions of Faces Each

For all twenty thousand proteins in your body to randomly fold and make one of you, it would take a stroke of luck equivalent to rolling twenty thousand dice at the same time and getting them all to land on 6! And note that each die is not made up of six faces but rather trillions of trillions of faces. Good luck with that!

I've tried to show you the hidden parts of the mathematics that get tucked away when the cases for the Big Bang, evolution, and natural selection are made. I've attempted to tell a story in which everything is designed to perfection and interoperates flawlessly with no dependence on luck, a story in which there's no chance and no trial and error, where everything behaves as expected and as per the everlasting cosmic equations set forth by the design. I can't prove that story with 100 percent certainty. But when it comes to our original problem statement it surely enjoys a significantly higher probability than the tale of randomness. This story *demands* the presence of a designer—which, unfortunately, is an entity that has been taken over and disfigured by religious institutions to the point where we would rather deny its existence than belong to the madness waged in its name.

Our universe is so complex that we often get lost in the details. Even Einstein admitted the limitations to our understanding: "[Looking at creation] we are in the position of a little child entering a huge library filled with books in many languages. The child knows someone must have written those books. It does not know how. It does not understand the languages in which they are written. The child dimly suspects a mysterious order in the arrangement of the books but doesn't know what it is. That, it seems to me, is the attitude of even the most intelligent human being toward God." [13]

Steve Jobs Who?

To remove the complexity, I learned to sum up my logic of the grand design in one simple story. Imagine if you had met me when I bought one of the very first iPhones. You are immensely impressed by it and so ask me where you can get one. How would you have felt if I'd told you this:

> I'm so sorry, but you can't purchase this technological marvel at a store. It happened through evolution. I found it in the 4.5-billion-year-old sand patch behind my garden.
>
>   I am the privileged recipient of a stroke of luck, a one-in-a-

gazillion chance, when the sand melted correctly to form the pure glass screen, which broke into the exact rectangular size, when a cat stepped on it at the exact right moment, to fit into an aluminum case. That pretty case was made out of a single block of refined aluminum that gathered over millennia from the garden soil and was then milled into shape by years of gentle sandstorms. The super-high-quality screen appeared inexplicably one morning, fell in place, and connected perfectly to the micro-electronics that self-formed from the silicon in the sand. Microphones and speakers resulted from insects eating their way through the device, and connections were made via refined copper found in an ancient artifact that was buried nearby. Entropy left this configuration untouched for years, until another incredible stroke of luck managed to bring together some chemicals, metal conductors, and packing to form a battery. A lightning bolt charged it as it fell in place during a minor earthquake. The rest of the aluminum formed as a result of the heat and finished the perfectly sealed case. Oh, and the software just wrote itself by random keystrokes on a randomly created computer that was found in our neighbor's garden around a year before.

Just Give It Time
It all comes from dust

It may have taken billions of years for this amazing miracle of probability to happen but our universe had enough time, so it did happen. This iPhone was created by randomness, and I don't care if saying this offends Steve Jobs because I really don't believe there ever really was a Steve Jobs anyway. Do you?

Why is it so hard to believe that a much more intelligent designer put together the machine that is you? Neither we nor our iPhones came into existence randomly.

Remember!
→ **There is a grand designer, and there was a Steve Jobs too.**

Back to Happiness

Over the years, I accepted the grand design as my foundation. I then spent many years questioning the consecutive layers one by one. In conclusion, I'm now convinced that there is a designer and that the designer sent us several messages in the core of several religions. I seek the wisdom in those messages and ignore the parts that got augmented with human interpretations, greed, and tradition.

This approach to life makes me happy because it helps me put to rest the debate inside my head, to know that I'm betting this short life on something I can prove with math, to know that I'm part of something bigger than this life, and that Ali is with a designer who's capable of creating such an awesome universe and is so much more capable of taking care of my beloved son than I ever have been.

When we started this conversation, I asked you to strip away the deeper layers of queries surrounding the question of design so that we could find an answer to the core problem statement. Now, to get back to the topic of happiness, we'll add just one of those layers back.

The Rules of the Game

The most crucial layer, causing us a lot of suffering, relates to our disagreement with the design. We humans, unlike the machines we make, constantly question the design. We think it should have been better. Our biggest disagreement with the designer, and the reason many reject the concept, is rooted in the disapproval of the way he behaves. (Please note that gender is a property of the physical world. I use the word *he* here for convenience and not out of gender bias.) The way the designer seems to work frequently misses the expectation side of our Happiness Equation, and that makes us unhappy. But should these actions be rightly attributed to the designer?

To start with, many of us disagree with his choice of "deputies" here on earth. The religious institutions, which claim ownership of the conduit to communicate with the designer, are really messing up. Most religions have become needlessly strict. They stray from their core premise; they advocate an exaggerated expectation of judgment; and they apply a "tax," while their own leaders frequently misbehave. None of this bothers me. I consider myself to be reasonably religious, despite the actions of the institution, because my loyalty rests firmly with the designer, not the *self-appointed* middlemen.

Beyond formal religion, many of what we perceive to be the designer's actions are hard to explain. Why is life so harsh? Why war, disease, death, destruction, hunger, filth, poverty, torture, crime, and corruption? Why do we suffer from natural disasters? Why did Ali have to be taken so young? If the designer is a loving, compassionate being, then he's clearly not running a tight show.

Well, I actually believe **the designer isn't running the show at all!** The equations he created do. Within that lies the beauty of the grand design and the ultimate truth—and happiness.

A tsunami is the result of seismic movements under the deep ocean that cause water waves to advance in the direction of land. There's no drama behind it. No intervention necessary. It's just the world unfolding in accordance with the laws of physics as per the design. When Audi produces a car, they make it so that when you put the gear in drive and push on the accelerator it moves forward. You may prefer to shout at it instead, but it will not move. It's just the way the design is. Audi will insist that you bring it in regularly for service and will drain the oil and replace it. This is not a defect; it's the design. You don't stand next to your car and complain about the process of oil change; you plug this process into your plans and expectations. Our Earth will occasionally seep volcanic lava; seismic shifts will cause earthquakes; and winters will be cold and harsh. When seven billion people are born, seven billion people will die. It is just the way it is. There's no drama; it's just a fact.

For an engineer, an equation represents the ultimate justice. An equation will always behave as expected. Depending on the values used as input, the output is absolutely knowable. Life and death, wealth and poverty, health and sickness, they just *happen*. Life is just what it is.

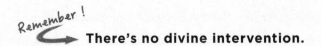

Remember!

There's no divine intervention.

Why does the design seem so harsh? I would have designed a kinder world. Why do snakes have to be venomous and so creepy? I would have created friendlier reptiles if I were the designer.

Well, that's a good point. But imagine, for a moment, a world without insects. Wouldn't that be wonderful? We could camp outdoors unbothered by all those creepy crawlers and mosquito bites! But don't get too excited. Without insects, there wouldn't be a place to go camping, as our earth would fill up with animal waste and dead plants in the absence of insects to break them down. Insects play a critical role as pollinators and food sources for other animals. Our food supply would be drastically reduced without them. Would removing insects make for a better design? No. The reason our world operates is because it behaves as one whole, one ecosystem. There really is nothing that's dispensable. If you ever doubt that, apply the eraser test here too. Feel free to erase any part of the universe that you don't like, along with all of its consequences, and see if you end up with a world that you like better.

Remember !

➤ **The design is just functional.**

And by the way, often what we complain about is our own doing. We wage wars that kill millions and blame an inattentive and cruel God. The world still provides for seven billion people, but our greed and waste leave a billion starving while another billion become obese. It's our behavior, yet we blame the design. We pollute, run nuclear experiments, acquire more wealth than we need, and use each other for our own selfish ends. And who do we blame for the damage? The big guy who is supposed to stop us. If we're driving and decide to take a sharp turn at 100 miles an hour, we should not blame the designer of the car for our reckless behavior, should we?

Remember !

➤ **The results of our own acts should not be blamed on the design.**

As you can imagine, Ali and I would not have had as much fun playing *Halo* if we just sat there to criticize the rules of the game. We just knew what they were, expected them to play out, and mastered the game within the limitations that were imposed.

And like a game, life too imposes some rules. Learning to master the game within them, instead of wishing they were different, can indeed take you where you need to be.

With this, I lovingly leave you to work on all the other layers. They are your puzzle. Solving them is part of your game.

Months have passed and I'm writing one last page that aims to nudge you firmly into your state of joy. If you come away with nothing else from this book, please understand this: there's no randomness in life. Our universe is the product of masterful design. The designer doesn't run the show; the equations he designed do. Focus on your Happiness Equation. It's the only one you can fully control. When you compare the events of your life to the expectations you set, please remember that what happens is what's supposed to happen. Instead of dreading the event, perhaps you should start by doubting the expectations you set because, harsh as that is sometimes:

Remember!
→ **Life always meets realistic expectations.**

Our universe is far too complex to predict. Surrendering oneself to a design that is beyond our ability to grasp is freeing. That freedom is joy. Try to shape your destiny and aspire to change the world for the better. Know, however, that your input into the equations of the grand design is limited. Arm yourself with committed acceptance in the instances when a billion other parameters tilt the result away from your expectation. Surrender and marvel at the brilliance of the design and you will have the ride of your life.

Very Important !

Accept the design.

I have shared with you hundreds of pages about what I believe to be illusions, blind spots, and truths, about what worked for me, knowing that not all of it will work for you. I could be wrong. So take what fits you and drop the rest, but know that one thing will always work: to find your peace and joy, reject every illusion and always seek what's real.

Very Important !

Live the truth. Finding joy truly is that simple!

Afterword

A Conversation with Ali

As I kissed the forehead of my departed son, an overwhelming feeling of peace gripped me. I couldn't explain it. Was it simply the removal of the anxiety Nibal and I felt as we stayed up all night without knowing what would happen to our beloved son? Was it relief from the pain we felt as we sat outside the intensive care unit knowing that our son was kept among the living only by life-support machines while we hoped the anesthesia kept his pain away? Or was it the peace of knowing he wouldn't have to endure a life suffering the collateral damage of the organ failures he sustained?

Those thoughts might have relieved us, but none of them was a good reason to feel *peaceful*. How was it that, in my heart, I wasn't raging about the series of preventable mistakes that took the life of my best friend?

I thought I'd gone crazy—but perhaps crazy was good. Crazy was peaceful. And that peaceful feeling helped me gather myself and walk outside the ICU to his fragile mother, so I could gently tell her the devastating news. Though he had already been officially declared dead, I tried to ease it a bit. "Nibal," I said, "it seems that Ali might not make it." Her reaction was even more surprising than mine. She said, "Take me to him." I pushed back; I wasn't sure that Nibal, with the aching heart of a loving mother, could stand seeing Ali in that state. But she smiled confidently and said, "I know he left, so take me to him. I want to say good-bye."

Ali looked so handsome, even in that state. He had trimmed his beard the day before and kept his curly hair short. His face looked relaxed, as peaceful as I've ever seen him. Nibal smiled a genuine smile, touched Ali's face, and said the most unexpected thing: "*Habibi,* you're finally home." That was clear. The initial feeling that warmed our heart was that he was okay—even better than okay. Ali was truly where he belonged.

We feel the same way today, but that wasn't always the case. As soon as we left the hospital, the gravity of the event set in and the peace we initially felt slipped away. It took everything we had to get it back.

For years my happiness model taught me to control the thoughts in my head. I could tell my brain to suspend a negative thought and get me a better one. With lots of practice, I had even been able to tell my brain to shut up completely and leave me in peace. The unexpected loss of Ali, though, threw me way out of balance. That crazy peace I felt at first quickly turned into malicious and aggressive thoughts. I was a total mess.

My tears wouldn't stop. The pain of missing Ali felt like a spear piercing my heart. The noise of my streaming thoughts deafened me. I felt I was going crazy—quite literally—especially when I began to hear the repeated sound of an unfamiliar, and weirdly cheerful, arpeggio as though from a guitar, playing monotonously over and over and over in my head. I just couldn't switch it off. It was insane.

It felt cruel. The one person I would normally seek out for advice on how to get out of such a spiral was exactly the person who was now gone. I desperately wanted to ask him, "Ali, how do I handle losing you?"

When I returned home after carrying his body to its final resting place, I collapsed into an exhausted sleep for a few short minutes and he showed up in a dream. I saw him sitting up from the operating table and turning toward me. He looked at me, smiled, then looked behind me and rushed by as if to hug someone who stood there, someone he clearly loved.

The sight of him made me jump out of bed. My heart was racing, but I felt peaceful for a moment. Then I remembered what had just happened. I cried. I realized how much I needed to see him again, even if in a dream, to talk it all through. For the following few days, I would close my eyes and imagine him walking toward me—perhaps on one of my long-haul flights—with that smile, the dreadlocks he had as a teenager, his black band T-shirt, and jeans.

I would jump out of my seat to hug him. *Ali, you're back, I missed you.* And, as he always did, he would say, Ezayak ya abuya. *How have you been, Dad?* Perhaps this time it wouldn't be the best question to ask. Because I would burst into tears and pour my heart out.

It's been hard, Ali. It's been very hard. We miss you and don't know how to live without you.

Tell me about it, Phat Hobbit. (That's what he jokingly called me since the day he became taller than I ☺.) *We have a long flight ahead and hours to chat.*

My brain is in hyperdrive, Ali. Nothing makes sense anymore. My thoughts are toxic: the doctor murdered my son; no one so young should die; life is unfair; there's no point in living another day; and a million more.

Illusions are taking me over to the point where I'm becoming almost delusional. The Illusion of Self makes me think that this was all because of me, that life was punishing me for something I'd done. My ego's hurting. I keep asking, Why was *my* son taken from *me*? The Illusion of Knowledge is pushing me to think that I should have known better than to take you to that hospital. Why didn't I pick a different medical team? I should have known better. The Illusion of Control is tearing me apart, shattering my faith in life. Why didn't I plan for this? What else didn't I plan for? The Illusion of Time is slowing down the clock, locking me into hours of tears, guilt, and anger about the past, as well as worry about the long future I have to live without you. The days feel excruciatingly long. I'm numb to the outside world, living inside a head that's filled with endless thoughts and emotions. And finally, my fear is overwhelming. Fear of what might happen to Aya, to Nibal, and what else this life might take away.

As I daydreamed about Ali and how he would work his magic and set my brain right—as he always did—all I could hear back was that annoying tune still playing. Then the first message came.

Will It Bring Ali Back?

News traveled fast. I received a call from a senior official in the Dubai government. He'd heard about what happened and promised that the medical negligence wouldn't be ignored. He said an investigation was already under way. He asked whether I would participate and if we were okay with an autopsy being performed on Ali's body. I turned to Nibal and asked her what she wanted to do. She said the wise words that anchored us in the ultimate truth: "Will it bring Ali back?"

It was like a lighthouse cutting through the fog. Nibal's question immediately refocused my thoughts. The truth was simple. The kindest man we ever knew left in peace. Nothing we could do—nothing— was going to bring him back. Any thought beyond that bare truth was malicious, worthless, and simply untrue.

Is That True?

From then on, the stray conversations in my head were balanced with sanity. Every time a malicious thought popped up, I could hear Ali's voice asking, *Is that true?*

The doctor murdered my son. *Is that true, Papa? What doctor wakes up in the morning and says, "Today is the day I'm going to kill someone and ruin my career"?*

No one should die at such a young age. *Is that true? Youngsters die by the thousands every hour of every day.*

My life will stop with yours. *Oh, is that true? Life stops for nobody. You're here until it's your time to go. You'd better behave with that in mind.*

This is the worst thing that could ever happen to me. *Is that really true, Papa? It could have been much worse, you know. I could have been diagnosed with a lingering cancer or drafted into the madness of the wars of the Middle East instead of leaving peacefully in my sleep.*

But I drove you there myself. I should have known better. *Is that true? How could you have known? You did what you thought was right, Papa. You wanted to ease my pain. You wanted me to recover. No one could have known this was going to be how things turned out. Knowledge is an illusion. Don't let it confuse you.*

I can't even live this way for a few days. It will torture me for years. *Is that true? You will live, and time will pass. The days will be long, and the years will be short. In no time at all you'll be looking back at this and wondering, "Has it really been that long since he left?" Life zooms by just as it has until today. Time is an illusion. Instead of thinking about the years to come, focus on now. Do the best you can. Make me proud. Live life one day at a time. When I was back in Boston we met only once a year, and it was okay. I'm just a bit farther now for a little longer. There's nothing to wait for. Time will take care of itself. You take care of you.*

But why is life doing this to *me*? It took away my son. *Is that true? With all due respect, Papa, I was never yours. I was always mine. This is my movie, and it's time for me to change the scene.*

There must be something I can do to change this. I've always been in control. *Haha! Is that even remotely true, Papa? No one's in control. We try the best we can. We take the right actions and keep the best attitude. The results are not up to us. Control is an illusion. What actions will you take today, and what attitude will you keep if you knew for certain that nothing would bring me back? Focus your thoughts on that.*

I know. Your death makes it all clear, but I'm so afraid. Will Nibal make it through? What will happen to Aya? *Fear too is an illusion, Papa. What will happen will happen, regardless of your fear. And in the end, we will all be fine. There's really nothing to fear.*

Fine? Are you really fine? Where are you now? Are you safe? Will I ever see you again?

The conversation in my head seemed endless. As his voice helped me to clarify one thought, another would pop up. Ali was still helping me, but my brain was working in hyperdrive to make me suffer, until I could hear Ali say, *Papa, we won't be together for a while. Isn't that enough? Why do you want to let your thoughts bring you years of suffering that won't change a thing? You know what I want you to do? I want you to be happy! Thoughts are an illusion. You can tell your brain what to think. Tell it to find the truth.*

The Truth

In Elisabeth Kübler-Ross's famous conception of grief, it all starts with denial. Then we move into bargaining, anger, and depression, before reaching acceptance.[1] Nibal and I were lucky to skip right past denial. There was no distortion of the event in our mind. The minute Ali's departure was announced, our approach was totally anchored in reality. *Ali left.* There was no point arguing that, and no way to bring him back. But that acceptance we felt quickly became distorted by thought, and suffering set in. We needed, as a family, to get that peace we felt at the ICU. Because of my research into happiness, I knew that the first and last place to find it was in the truth.

In this case, the truth was simple. Ali had a full life. He was intensely present. He lived every minute, and he was always happy. Somehow—without my malicious thoughts—I felt okay.

I began to shift my focus away from what Ali no longer is to who he was. Ali was a kind guest who brought light and happiness to our home. But guests are not expected to stay forever. He didn't leave too soon. I thought back to the day he blessed us with his presence, twenty-one full years earlier, and thought how fast the years had zoomed by. Even if I had twenty-one more with him, they would've zoomed by just as fast. Instead of thinking about losing him, I learned to think about the beautiful truth: we *had* him. For all those years he brought joy to our life. Instead of feeling sad that he left, for the first time I felt happy that he came to visit in the first place.

And then I relied on a strong belief in the two truths that are hardest to accept: death and design. I thought back to the research I had done in my youth into the concept of design, into the encrypted, sometimes distorted messages frequently found in religious teachings, and I realized that what was common among all faiths was this simple core message: Death is not the end, and those who are kind in this life will be just fine in the next round of the game. Death of our physical form is very real, but death is not the end. The real me, the real you, and the real Ali never really die. Sooner or later, I will join Ali to explore the other side. The designer is generous and kind, despite events in this physical existence that might sometimes lead us to believe otherwise. He will take better care of Ali than I ever did.

No, there's no way I can prove this with total certainty, but isn't that the nature of all knowledge? And when my brain attempted to take over, when my doubt and cynicism and chatter were ascendant, I remembered my golden rule of happiness: When you choose between two thoughts and can prove neither with total certainty, choose the one that makes you happy. What could be simpler?

I choose to be happy. Ali's fine. He's on the next level of his game.

His Last Words

As these positive thoughts started to dominate my head, I could at last think clearly. I realized that the conversation I'd been hoping to have with Ali had started, and ended, even before he left.

During his last few weeks, Ali asked almost everyone he met, "What happens to us when we die?" He rarely spoke of anything else. It was as if he were preparing for a journey that he knew he was going to take. He was curious. He asked the question and listened attentively. He didn't judge or debate. A few days before his unexpected departure, in a last conversation about the topic, he shared his own view: "Well, I guess we will only know when we get there. But I am optimistic!" He was ready. He found his peace even before he left. **That made me happy.**

One morning, a week before he left, he told his sister that he'd had a dream. In his dream he was everywhere and part of everyone. He said that no words could describe how he felt but that he didn't want to be caged in his physical body anymore. When he left, thousands of people in every corner of the globe were touched by his story. Even today, many people tell me they love Ali though they never met him. This book is bringing him to even more people. He's everywhere and part of everyone. His dream came true. **That makes me happy.**

A few days later Ali spent time to dictate his last advice the way a wise grandfather might. He told Nibal, Aya, and me how much he loved us, and then asked us each what needed to be done to truly face life. He said, "I don't know why I am saying this unsolicited, but I feel compelled to." He filled our hearts with love, showered each of us with kind words, and then offered his advice. He asked his mother to simply "be happy" and to be open to exploring life like Walter Mitty. (Watch the movie.) He asked his sister to "be aware" and to find her real beautiful self. And he asked me to never stop working. He said, "You're doing so well, Papa, and you're making a difference. Your work here is not done." Those words changed my life and made me who I am today. He told me what I needed to do to survive losing him, and **that makes me happy**.

He then smiled peacefully. His face reflected a sense of satisfaction, one that could only mean *But now my life's work is done*. In his usual musical way he then said, "Okay, that's it. I have nothing more to say." Those were his last words for the day. He went back to his usual quiet self—even quieter with every passing day.

As his last days approached, he spoke little, slept a lot, and ate almost nothing at all. It was as if he had used up his budget of the things we do while living.

His very last words to me came unspoken. They have kept me strong every day since. Ali had only one regret in life: a tattoo he got as a teen and hid from me for years. He knew I would support his getting it, but he felt guilty because he had used my money (though it was his allowance) without asking my permission. For years he waited for an appropriate time to tell me. It was the only secret he ever kept. He did tell his mother, though, and she, of course, told me. I didn't want to bring it up until he was ready. For some reason,

Nibal reminded me of it when we got to the hospital, so it was fresh on my mind.

When he was on the way to the operating room he sat up, and I saw the tattoo for the first time. Aloud I said, "I approve, *ya habibi*." I hope he heard me so that he knew there was no reason for his regret. But either way his peace was complete. He did tell me, even if unintentionally. He was guilt-free and **that made me very happy**.

His tattoo contained his last words to me. It was the ultimate statement of truth.

The gravity of the battle means nothing to those at peace.

Thanks for reminding me, Ali, and thanks for the wonderful conversation.

You may think I was just looking for anything to hold on to to stay positive. And I was. There's absolutely nothing wrong with that. Suffering is a choice and I chose not to suffer. By then the thoughts were gone. Yet, I continued to see positively in my dreams.

The night of his memorial, right after everyone left, I slept and, in a dream, saw him standing in the crowd. He was crossing his arms and looking around with a big smile on his face and loving eyes for all the people who had come to pay their respects. He seemed happy and proud of the positive happy energy that filled the place.

A few nights after I started writing this book, I had another dream. He was dancing, spinning around, laughing, and waving his arms in the air. He sang in a cheerful tune, *But I'm proud of everyone, proud of everyone*. I was still feeling the pain of losing him, but I was happy. Not only because of my dreams but also because I had already deciphered a much clearer message from my beloved son. But even then one thing was still confusing me.

Remember that crazy tune that kept playing over and over in my head after his departure? I tried hard to silence it, but then I realized that it must be there for a reason. It took me a while to realize it too was a message. The conversation I'd been seeking with Ali was staring me in the face all along, and I missed it. It was encoded in that tune.

Portal

Ali was serious about video games, and he found messages and philosophy in the way games simulated life. He was also a talented musician. If he were to send me a message, I thought, how else would he do it other than through music and games? When I concentrated, I remembered that I had previously heard that tune—but only once, at a concert I attended with Ali years before. It was the theme song, written by Jonathan Coulton and performed by Ellen McLain, for a game, appropriately called Portal, that we both loved. The song played along with the credits at the end of the game, but because I hadn't finished the game (I have now, of course), I had heard it only at that concert.

Portal is about an evil computer, Glados, that pretends to be your friend as she guides you through some experiments in Aperture Science Laboratories. You can't help but love Glados. Like a voice inside your head, she seems to be helpful. She tells you what to do, and what she says usually sounds like good advice. She motivates you and promises to give you cake if you perform the difficult tasks. But as you get deeper into the lab, you find frantic graffiti on the walls: "The cake is a lie." And at the end of every level, Glados fails to deliver on the promise. But even though you don't get any cake, you still love her because she's funny—*really funny*—so you forgive her.

Midway through the game, however, you discover that she—like the little voice in your head—is lying. She's trying to hurt you. And you realize that, like all of life's

The cake is a lie
The cake is a lie
The cake is a lie

illusions, the cake *is* a lie. You win the game only when you stop listening to her voice and focus on the reality of the gameplay. Sound familiar?

Throughout the game you use a portal gun, which is a weapon that doesn't kill but rather helps you shoot portals through the physical boundaries of the lab to take you from where you are to where you want to be. When I think about it, there's no better game to summarize the situation of Ali's departure through an unexpected portal.

When I remembered where the song came from, I looked for it on YouTube, and the top search result was one that included the lyrics. Almost every word had the same witty lighthearted tone in which Ali always spoke. It starts with the sound of distant white noise, as though you're tuning a radio to receive a broadcast message from far away. After a few seconds, along with a soothing guitar arpeggio, a tender synthesized voice that sounds as if it comes from another dimension starts to sing:

This was a triumph
I'm making a note here: **huge success**
It's hard to overstate my satisfaction.[2]

I shivered. I paused the music and couldn't hold back my tears. If this was a message from my son, then it couldn't be clearer. He was okay. No, more than okay. He was glorious. He looked back at the whole game of his life and death and concluded—in one word—that it was a *triumph*. The song went on to remind me of the mission he gave me:

Aperture Science
We do what we must
because we can
For the good of all of us.
Except the ones who are dead.

It echoed his advice to me: *Never stop working, Papa. Make a difference. Not for any reason, but just because you can. There is nothing you can do for the ones who are dead, but you can for the living.* He was saying that just days before he left. It was my assigned mission. The song continues:

But there's **no sense crying** *over every* **mistake.**
You just keep on trying till you run out of cake. . . .

Don't get stuck crying about the human error that took my life. Move on till your time in this life is done. Focus on life and do good. That's how you spend the rest of your life. Onward.

I'm not even angry.
I'm being so sincere right now.
Even though you broke my heart.
And killed me. . . .

My wonderful son has forgiven—as he always did. He knows that good will come out of this somehow, and he's happy for us to go through this experience.

> *Go ahead and leave me.*
> *I think* **I prefer to stay inside**.
> *Maybe you'll find someone else to help you.*

Go ahead and leave me?! Once again I cried, but the message was clear. *Move on, Papa. You know what to do. There will be others who will help you on your mission—hopefully all those readers that will help spread the message.* I found this part of the message hard. My heart was aching. I wanted my son by my side but could not have him. Then at the end of the song he told me why:

> **. . . And believe me I am still alive.**
> *I'm doing Science and I'm still alive.*
> **I feel FANTASTIC and I'm still alive.**
> *While you're dying I'll be still alive.*
> *And when you're dead I will be still alive.*
> **still alive**
> **still alive**

I know you are, *ya habibi*. I am sure you're happy wherever you are—chatting with the most interesting people who ever lived.

I too will finish my work here and find my portal. We all do someday. Through it all, I'll miss you, but I promise to be happy as you would want me to be. I promise to make you proud. And I'll always be thankful that you showed me the way.

Thank you for reading my story and considering some of my views on life. I hope that you find your happiness too and truly hope to meet you someday. Meanwhile please write and teach me how applying some of those concepts has worked for you.

In his last conversation about death, when Ali told his friend he was optimistic, he stated his last wish: *All I want when I get to the other side is to go to the highest place and see the face of the one who made this awesome universe.*

Please pray for Ali that his wish will come true.

Acknowledgments

Solve for Happy would not have been possible with out the incredible support the whole North Star Way team granted me. Michele Martin, I adore your vision, assertiveness, and coaching. Thanks for signing up to the mission. Diana Ventimiglia, you truly turned this upside down. Always with a smile, you got us were we always needed to be. I had so much fun, it did not feel like work. This book is leaps better than the one I came to you with, Michele and Diana. Thank you.

My journey introduced me to Michael Carlisle, my agent and now friend for life. You believed in my mission and guided me lovingly. I can never pay you back.

Nibal, umm Ali, thanks for all the years, wisdom, friendship, and love. Every thought in this book was born from a conversation with you. I would have been a very different person had it not been for you.

Aya, the sunshine of my life, I love you and love our conversations. I've learned so much from you my daughter. Shine on you crazy diamond.

Ummy, Amira Wahby, you're the best. Thanks for letting me read when I was so young and letting me explore when I became older. And thanks for always being there.

Thanks Carole Tonkinson for you vote of confidence so early in this journey and for helping me build the foundation of this book.

Peter Guzzardi, I wish we spent more time together. Thanks for all your experience, patience and encouragement.

William Callahan, you're a dynamo. Insightful, eloquent, driven, and fast. Man, you're fast.

Rick Horgan, what you taught me is priceless. To you, I have the utmost gratitude and respect.

William Patrick, thanks for your services.

Tento, I thought I was done learning till you said stop judging, predicting, and say it as it is. That one of many tips helped shape the way I looked at things. Thanks for all of them.

Thanks to Ellis and the Chartwell Speakers team for getting me in front of thousands.

Gracias Marcella Gomez for spreading the mission in Latin America.

Jennifer Aaker you literally hugged *Solve for Happy*. Sharing the concept with Stanford helped me dig into some of the world's smartest minds.

Betty Lin thanks for helping me take it all the way across the world to Hong Kong and Emily Ma, thanks for everything in between.

Halfway through this journey, I posted an early version of *Solve for Happy* online. Hundreds of early readers reviewed it. They argued, debated, shared research, and even edited the words themselves. Tens of thousands of comments and changes made this a book as close to being written by the readers as books get. Anne, Ossama, Karla, Lori-Ann, Gulnara, George, May, Alix, Nader, Emily, Maysam, Emel, Eslam, Hana, Agnieszka, Yee Hui, Astuti, Jenni, Dina, Samaa, Aurore, Gladys Karina, Karishma, Evan, Angela, Lamia, Nikesh, Tracy, Viviana, and all the others who so generously contributed, thank you. To each of you, I'm eternally grateful.

Thanks to all the authors and thought leaders, whose quotes and books I have referenced and whose wisdom lit my way.

Thanks to all the tough times, that forced me to search, research, and introspect. I would not erase a thing.

Thank to all of you that I still have not yet met who will sign up, volunteer and help us reach the mission. I can't reach #10millionhappy without you.

And thank you Ali. For everything that you showed me, the love that overwhelmed me, and for giving me a reason to write. I love you, son. Stay happy till we meet when my work here is done.

Notes

1. Setting Up the Equation

1. Ed Diener and Richard Easterlin, "Rising Income and the Subjective Well-Being of Nations," *Journal of Personality and Social Psychology* (2013), https://www.apa.org/pubs/journals/releases/psp-104-2-267.pdf.
2. Andrew J. Oswald, Eugenio Proto, and Daniel Sgroi, "Happiness and Productivity," Warwick Social Sciences, working paper, February 10, 2014, https://www2.warwick.ac.uk/fac/soc/economics/staff/eproto/workingpapers/happinessproductivity.pdf.
3. Malcolm Gladwell, *Outliers: The Story of Success* (Little, Brown, 2008).
4. Daniel Gilbert, *Stumbling on Happiness* (Vintage, 2007).

2. 6-7-5

1. Mihaly Csikszentmihalyi, *Flow: The Psychology of Optimal Experience* (Harper Perennial Modern Classic, 2008).

3. That Little Voice in Your Head

1. Eckhart Tolle, *A New Earth: Awakening to Your Life's Purpose* (Penguin, 2008).
2. Gartner, "Gartner Says 6.4 Billion Connected 'Things' Will Be in Use in 2016, Up 30 Percent from 2015," press release, November 10, 2015, http://www.gartner.com/newsroom/id/3165317.
3. Daniel Kahneman, *Thinking, Fast and Slow* (Farrar, Straus & Giroux, 2013).
4. Bhavin R. Sheth, Simone Sandkühler, and Joydeep Bhattacharya, "Posterior Beta and Anterior Gamma Oscillations Predict Cognitive Insight," *Journal*

of Cognitive Neuroscience 21.7 (2009), http://www.mitpressjournals.org/doi
/abs/10.1162/jocn.2009.21069#.Van3LhOqpTI.

5. Norman A. S. Farb et al., "Attending to the Present: Mindfulness Meditation
Reveals Distinct Neural Modes of Self-Reference," *Social Cognitive and Affective
Neuroscience* 2.4 (2007), http://scan.oxfordjournals.org/content/2/4/313.full.

4. Who Are You?

1. *The New York Public Library's Science Desk Reference* (Stonesong Press, 1995).
2. Nicholas Wade, "Your Body Is Younger Than You Think," *New York Times*,
August 2, 2005, http://www.nytimes.com/2005/08/02/science/your-body-is
-younger-than-you-think.html?_r=0.

5. What You Know

1. Donald Rumsfeld, U.S. Department of Defense news briefing, February 12,
2002, *Wikiquote*, https://en.wikiquote.org/wiki/Donald_Rumsfeld.

7. Houston, We Have a Problem

1. Nassim Nicholas Taleb, *The Black Swan: The Impact of the Highly Improbable*
(Random House, 2010).
2. "Butterfly Effect," n.d. https://en.wikipedia.org/wiki/Butterfly_effect.

8. Might as Well Jump

1. "John B. Watson," n.d. https://en.wikipedia.org/wiki/John_B._Watson.
2. "Pain tolerance," n.d. https://en.wikipedia.org/wiki/Pain_tolerance.

9. Is It True?

1. Mihaly Csikszentmihalyi, *Flow: The Psychology of Optimal Experience* (Harper
Perennial Modern Classic, 2008).
2. Raj Raghunathan and colleagues, *If You're So Smart, Why Aren't You Happy?*
(Portfolio, 2016).
3. Chopra, Deepak, "Why Meditate," *Deepak Chopra,* https://www.deepak
chopra.com/blog/article/470/.
4. Roy F. Baumeister, Ellen Bratslavsky, Catrin Finkenauer, and Kathleen D.
Vohs, "Bad Is Stronger Than Good," *Review of General Psychology* 5.4 (2001),
http://dare.ubvu.vu.nl/bitstream/handle/1871/17432/Baumeister_Review
?sequence=2.

5. Felicia Pratto and Oliver P. John, "Automatic Vigilance: The Attention-Grabbing Power of Negative Social Information," *Journal of Personality and Social Psychology* 61.3 (1991), http://people.uncw.edu/hakanr/documents/AutoVigilanceforneg.pdf.

6. David L. Thomas and Ed Diener, "Memory Accuracy in the Recall of Emotions, *Journal of Personality and Social Psychology* 59.2 (1990), http://psycnet.apa.org/psycinfo/1991-00334-001.

7. Tugend, Alina. "Praise Is Fleeting, but Brickbats We Recall," *New York Times,* March 23, 2012. http://www.nytimes.com/2012/03/24/your-money/why-people-remember-negative-events-more-than-positive-ones.html.

8. Rick Hanson, *Just One Thing: Developing a Buddha Brain One Simple Practice at a Time* (New Harbinger Publications, 2011).

9. Christopher Chabris and Daniel Simons, "The Original Selective Attention Task," *The Invisible Gorilla,* http://www.theinvisiblegorilla.com/videos.html.

10. Vilayanur S. Ramachandran and Diane Rogers-Ramachandran, "Extreme Function: Why Our Brains Respond So Intensely to Exaggerated Characteristics, *Scientific American*, July 1, 2010, http://www.scientificamerican.com/article/carried-to-extremes/.

11. Daniel Kahneman, *Thinking, Fast and Slow* (Farrar, Straus & Giroux, 2013).

12. Dan Cray, "How We Confuse Real Risks with Exaggerated Ones," *Time*, November 29, 2006, http://content.time.com/time/health/article/0,8599,1564144,00.html.

13. Shawn Achor, "The happy secret to better work," *TED*, February 2012, www.ted.com/talks/shawn_achor_the_happy_secret_to_better_work.

Part Four

1. Stanley, Jan B., "Arianna Huffington is Redefining Success," *livehappy*, May 21, 2015, www.livehappy.com/lifestyle/people/arianna_huffington_redefining_success.

10. Right Here, Right Now

1. Matt Killingsworth, "Want to be happier? Stay in the moment," *TED*, November 2011, https://www.ted.com/talks/matt_killingsworth_want_to_be_happier_stay_in_the_moment.

11. The Pendulum Swing

1. Robert A. Emmons and Michael E. McCullough, "Counting Blessings versus Burdens: An Experimental Investigation of Gratitude and Subjective Well-Being in Daily Life," *Journal of Personality and Social Psychology* 84.2 (2003), http://www.stybelpeabody.com/newsite/pdf/gratitude.pdf.
2. "In Praise of Gratitude," Harvard Mental Health Letter, November 2011, www.health.harvard.edu/newsletter_article/in-praise-of-gratitude.

12. Love Is All You Need

1. Elizabeth W. Dunn, Lara B. Aknin, and Michael I. Norton, "Spending Money on Others Promotes Happiness," *Greater Good*, n.d., http://greatergood.berke ley.edu/images/application_uploads/norton-spendingmoney.pdf.

13. L.I.P.

1. World Health Organization, Global Health Observatory (GHO) Data, http://www.who.int/gho/en/.
2. Peter Saul, "Let's Talk about Dying," *TED*, November 2011, https://www.ted .com/talks/peter_saul_let_s_talk_about_dying?language=en.
3. An observing life-form could have been one of ours or that of an extraterrestrial residing outside our own universe. Unless that extraterrestrial life created a new set of physical laws that are different for our universe than its own, it would suffer the same challenge of creation on its side and need a life-form to observe its universe at inception. And that life form would need yet another life-form to observe it into being. And so on. Alternatively, the life-form could be the Designer of all life, a topic for the next chapter.
4. Anita Moorjani, "Dying to Be Me!," *TEDx*, December 11, 2013, http://tedx talks.ted.com/video/Dying-to-be-me-Anita-Moorjani-a.

14. Who Made Who?

1. George Carlin, https://youtube.com/8r-e2NDSTuE.
2. Brian Greene, *The Fabric of the Cosmos: Space, Time, and Texture of Reality* (Vintage, 2005).
3. Sentences per second (S) = (9 *words* / 220 *words/minute*) $\star$ 60 *seconds/minute*.
4. Years to finish (Y) = (C) x (S) / $(60\star60\star24\star365)$ *seconds/year*. Number of possible configurations (C) = 27 *possibilities* $\wedge$ 56 *positions*.
5. Multiples of the age of Earth = (Y) / 4.5 $\star$ $10^{\wedge}9$ *years*.

6. Attempts to write *War and Peace* randomly = 27 *possibilities* ^ (580,000 *words* ★ 6 *letters/word*).

7. Typing speed needed to finish in time = *(C)* ★ *9 words/sentence* / (60 ★ 24 ★ 365) *minutes/year* ★ 4.5 ★ 10^9 *years*.

8. World Registry of Marine Species, http://www.marinespecies.org/.

9. Camilo Mora et al., "How Many Species Are There on Earth and in the Ocean?," *PLOS Biology*, August 23, 2011, http://journals.plos.org/plosbiology /article?id=10.1371/journal.pbio.1001127.

10. Attempts as a multiple of human population *(M)* = *(C)* / 7 ★ 10^9 *humans*.

11. Skeletons per square meter = *(M)* / 1.49 ★ 10^14 *surface area of the land mass*.

12. Humans to be created per month = *(C)* / 2 *million years* / 12 *months*.

13. George Sylvester Viereck, *Glimpses of the Great* (Macaulay, 1930), http://www .thinknice.com/albert-einstein-quotes/.

Afterword

1. Elisabeth Kübler-Ross, *On Grief and Grieving, Finding the Meaning of Grief Through Five Stages of Loss* (Scribner, 2007).

2. Jonathan Coulton, "Still Alive," YouTube, https://www.youtube.com/watch ?v=Y6ljFaKRTrI&spfreload=10.